清洁发展机制与中国清洁发展机制基金

中国清洁发展机制基金管理中心　编著

经济科学出版社

图书在版编目（CIP）数据

清洁发展机制与中国清洁发展机制基金／中国清洁发展机制基金管理中心编著．—北京：经济科学出版社，2011.12
ISBN 978－7－5141－1450－8

Ⅰ．①清…　Ⅱ．①中…　Ⅲ．①无污染工艺－发展－研究－中国 ②无污染工艺－发展－专用基金－研究－中国　Ⅳ．①X383 ②F832.21

中国版本图书馆 CIP 数据核字（2011）第 277399 号

责任编辑：凌　敏
责任校对：徐领柱
技术编辑：李　鹏

清洁发展机制与中国清洁发展机制基金
中国清洁发展机制基金管理中心　编著
经济科学出版社出版、发行　新华书店经销
社址：北京市海淀区阜成路甲 28 号　邮编：100142
教材分社电话：88191343　发行部电话：88191540
网址：www.esp.com.cn
电子邮件：lingmin@esp.com.cn
北京中科印刷有限公司印装
787×1092　16 开　15.25 印张　260000 字
2012 年 8 月第 1 版　2012 年 8 月第 1 次印刷
ISBN 978－7－5141－1450－8　定价：52.00 元
（图书出现印装问题，本社负责调换）

编委会名单

序一

把握机遇　积极开拓
创新气候变化融资机制

气候变化融资机制，一直是全球气候变化谈判的焦点问题之一和应对气候变化行动的关键举措之一。为此，各国都在努力探索、创新融资机制。中国的行动受到世界的高度关注。

《京都议定书》下的清洁发展机制，是全球创新应对气候变化融资机制的重要成果之一，历经数年发展与实践，已充分证明其在促进发展中国家可持续发展和全球共同应对气候变化方面的有效性，实现了发达国家与发展中国家的互利共赢。

中国政府高度重视应对气候变化问题，采取了一系列务实行动，积极参与包括清洁发展机制在内的国际合作，并努力探索适合中国国情的应对气候变化融资机制，以全面、迅速、有效地支持国内应对气候变化行动。成立中国清洁发展机制基金，就是这些务实行动之一，也是中国政府在创新应对气候变化融资方面对世界的贡献。

中国政府把中国清洁发展机制项目产生的温室气体减排量交易额中属于国家的部分，通过中国清洁发展机制基金，集中使用，专门用于支持国家应对气候变化战略、政策的制定与实施，支持国家减缓和适应气候变化行动，从而把清洁发展机制这一项目层级的国际合作机制，在中国提升为国家层级的国际合作机制，放大了清洁发展机制的作用，弥补了清洁发展机制原设计中的不足。同时，通过资金运作，中国清洁发展机制基金可引导和撬动社会各方的资金资源，投入国内应对气候变化工作，推动国家应对气候变化战略的实施。

目前，全球应对气候变化谈判进入了关键时期。虽然德班气候变化大会上决定《京都议定书》第二承诺期将继续，但有关核心问题仍待解决。在《京都议定书》第一承诺期（2008~2012年）结束之前，尽快解决这些核心问题，实现第一承诺期和第二承诺期的无缝对接，已成当务之急。而资金与技术问题依旧是后续谈判的重中之重。

受到全球高度注视的中国所制定的《国民经济与社会发展第十二个五年规划纲要》，把应对气候变化作为转变经济发展方式的重要抓手，并制定了明确的节能减排约束性目标；《“十二五”控制温室气体排放工作方案》就如何实现“十二五”期间节能减排目标提出了详细的实施措施。这表明我国在推动绿色低碳发展的工作迈上了一个新的重要台阶，也为财政支持应对气候变化工作指明了方向，为中国清洁发展机制基金提供了广阔的发展空间。

《清洁发展机制与中国清洁发展机制基金》一书全面阐述了全球和中国清洁发展机制项目发展状况，归纳总结了中国开展清洁发展机制项目的成功经验，介绍了中国清洁发展机制基金的成立背景、治理架构、发展战略、业务规划及其运作模式。该书对全球气候变化谈判中的资金机制议题谈判具有一定的参考价值；对我国进一步创新气候变化融资机制，具有借鉴意义；也为其他国家尤其是发展中国家，结合本国情况，创新气候变化融资机制，提供了一个范例。

希望中国清洁发展机制基金为全球应对气候变化事业做出积极贡献。

朱光耀

序二

发展清洁能源　增强创新气候融资的可持续性

为促进亚太地区消除贫困，亚洲开发银行一直积极支持各成员国追求环境友好的可持续发展。在过去二十年里，亚太地区保持了经济高速增长，但这一地区如不积极努力减缓和适应气候变化带来的不利影响，消除贫困和改善生活质量的努力将难以持续。作为唯一坐落于亚太地区的多边发展银行，亚洲开发银行在推动创新融资支持清洁能源项目和减缓与适应气候变化行动中，扮演了重要角色。

清洁发展机制是《京都议定书》下引入的，通过核证减排量（CERs）促进发展中国家清洁能源项目实施的一个重要机制。它帮助发展中国家获得技术和资金，更重要的是推广低碳发展的理念。清洁发展机制已被公认为是发达国家与发展中国家在碳减排方面的双赢机制。到2011年底，约有超过70个国家的3,900个项目已正式注册为清洁发展机制项目，每年将撬动超过400亿美元的投资。

中国一直是清洁发展机制的积极参与者。自2005年以来，已有大约1,800个项目注册成功，在全球已注册项目中占有最大份额。对清洁发展机制的成功应用，已帮助中国有效利用国际资源推进其气候变化相关活动。这不仅帮助中国从一开始便利用更清洁技术应对气候变化，还使低碳发展方式在一个快速发展的国家成为主流。由清洁发展机制项目产生的原动力已帮助中国成为清洁能源投资与制造领域的领头羊。

长期以来，气候变化融资的可持续性一直是气候变化政策讨论的焦点问题。中国清洁发展机制基金的建立是中国政府有效利用其清洁发展机制资源的创举之一。中国清洁发展机制基金通过征收一定数量的核证减排量交易收入筹集资金，用于加强国家应对气候变化研究，并撬动更多的公共和私营部门资金以提升减缓应对气候变化的努力。它是一个增强清洁能源项目气候融资可持续性的创新融资平台。

亚洲开发银行一直与中国政府在各种减缓和适应项目中保持密切合作。根据中国政府需求，亚洲开发银行自2006年以来，先后向中国清洁发展机制基金提供了两个技术援助赠款项目，支持其建立与能力建设。我很高兴地看到，中国清洁发展机制基金已经取得了很多令人鼓舞的成绩。我愿重复亚洲开发银行行长黑田东彦在2007年中国清洁发展机制基金成立仪式上的话，“中国清洁发展机制基金是一个首创之举，它利用国际碳市场的收入消除内部各种障碍，发展低碳经济”。

亚洲开发银行很高兴能资助《清洁发展机制与中国清洁发展机制基金》一书的出版。这本书简要回顾了《京都议定书》与清洁发展机制的发展历程，详细介绍了中国清洁发展机制基金的成立与宗旨，介绍了中国开发和实施清洁发展机制项目的成功经验。随着应对气候变化全球努力的加强，对知识创新和最佳实践分享的需求显得更为重要。毋庸置疑，这本书可以与其他发展中国家分享中国清洁发展机制发展的成功经验，并帮助国际组织强化对应对气候变化国际合作机制的理解。

当前，全球应对气候变化的努力距离有效应对气候变化的需求仍相距甚远。我们可以回顾一下清洁发展机制项目对中国快速增加可再生能源和清洁能源比重，以及在清洁能源竞争中建立领导地位的刺激作用。我们当然更希望中国清洁发展机制的成功案例能在本地区被广泛复制。

亚洲开发银行将继续与中国政府保持密切合作，推进其减缓和适应气候变化项目。我希望中国清洁发展机制基金有一个美好的未来，并在全球气候变化合作中发挥积极作用。

亚洲开发银行副行长

史蒂芬・P・格罗夫

前　　言

清洁发展机制，是《联合国气候变化框架公约》下的《京都议定书》规定的三种灵活履约机制之一，是目前包括我国在内的发展中国家参与碳交易活动最多的形式。

我国的清洁发展机制项目自2005年以来快速发展，目前项目数量及交易金额均占全球的50%以上。同时，我国基于清洁发展机制碳交易而建立的中国清洁发展机制基金，也迅速成长壮大，为国家应对气候变化和低碳发展提供了及时、有力的支持。这一全球独创性的融资模式得到了国际社会的广泛关注和好评。

在中国清洁发展机制基金管理中心执行亚洲开发银行支持的技术援助项目过程中，亚洲开发银行的项目官员认为：我国在清洁发展机制项目开发与实施、中国清洁发展机制基金的建立与运行，以及它们在支持中国开展应对气候变化和低碳发展所发挥的作用方面，成就斐然，积累了丰富的经验。这些实践、成果和经验非常值得总结，并与其他发展中国家共同分享。

感谢亚洲开发银行对本书出版给予的支持。感谢亚洲开发银行高级气候变化专家杨红亮，中国清洁发展机制基金管理中心孙玉清、王宁、温刚、涂毅、黎蕾等同事在编写过程中提出的宝贵意见和建议。此外，中国清洁发展机制基金管理中心田晨、中国社会科学院国有资产经营管理有限责任公司郑黎黎、新华通讯社霍焱、联想（北京）有限公司刘珍等承担了本书英文版的翻译与校对工作。

在本书的编写和出版过程中，经济科学出版社做了大量细致的工作，在此一并感谢。

虽然在本书编写过程中，编写人员努力将相关成果和经验尽可能完整、准确地呈献给读者，但囿于编者的视野、知识所限，书中定然有疏漏和能力不及之处，在此恳请读者批评指正。

编写组

2012年6月

目　　录

上　篇　清洁发展机制

下 篇 中国清洁发展机制基金

上　篇

清洁发展机制

第1章

气候变化与《联合国气候变化框架公约》

人类活动排放的温室气体已导致全球气候变化。全球气候变化对人类、社会、生态环境等的影响正日益凸显，采取有效措施减缓和适应全球气候变化，已成为当今世界共同面对的最严峻挑战之一。

1.1 气候变化

1.1.1 气候变化的成因

人类赖以生存的地球表面一方面接受来自太阳的短波辐射，同时向太空发射长波辐射。太阳的短波辐射几乎可以无阻拦地透过大气层到达地球表面，但地球表面的长波辐射会被地球表面围绕的温室气体吸收，形成温室效应，使地表与低层大气温度升高，并稳定在适宜人居及地球生物增长的近恒温环境。在过去几千年内，全球的温室气体浓度基本处于稳定状态或增加极为缓慢，全球温室效应达到平衡，使得地球表面温度接近于恒定。

但自1750年，特别是工业革命以来，由于发达国家工业化进程的快速推进，人类活动排放的温室气体大幅增加。根据政府间气候变化专门委员会（Intergovernmental Panel on Climate Change，IPCC）2007年发布的第四次评估报告显示，自工业革命以来，特别是在1970～2004年期间，由人类活动引起的全球温室气体排放增加了70%，其中，二氧化碳这一最重要人为温室气体的排放量增加了大约80%。2005年大气中二氧化碳（379ppm[1]）和甲烷

1 ppm：parts per million，百万分率。

（1,774ppb[1]）的浓度远远超过了过去650,000年的自然变化的范围[2]。

根据政府间气候变化专门委员会预测，若到2030年及以后，在全球混合能源结构配置中化石燃料仍保持其主导地位，全球温室气体排放量在2000～2030年期间增加25%～90%，则未来20年将以每10年大约升高0.2℃的速度变暖。即使温室气体和气溶胶的浓度稳定在2000年的水平不变，预计也会以每年大约0.1℃的速度进一步变暖[3]。全球气候变化形势严峻。

1.1.2 气候变化的危害

若温室气体继续以当前或高于当前的速率排放，将会引起全球的进一步变暖，并给全球带来严重的灾难和影响。具体包括[4]：

导致海平面的上升：根据政府间气候变化专门委员会的评估，若全球平均温度比工业化之前高1.9～4.6℃的情况持续千年，则将导致格陵兰冰盖的完全消失，并造成海平面比125,000年前上升约7米。

导致自然灾害增加：包括将导致北半球冬季缩短，并且更冷更湿，而夏季则变长且更干更热；亚热带地区将更干燥，而热带地区则更湿；积雪面积缩小，大部分多年冻土区域的融化深度增加，海冰面积退缩；极端气候事件如热浪、强降水等的频率增加；由于气温增高，水汽蒸发加速，导致全球每年雨量减少，全球各地区降水形态改变；高纬度地区降水增加，大部分亚热带陆地区域降水减少等。

对某些系统、行业和区域产生影响：对陆地、海洋和海岸带的生态系统，对中纬度一些干旱地区和干旱的热带地区以及依靠冰雪融化的地区的水资源，面临水资源减少的低纬度地区的农业，地势低洼的沿海系统，低适应能力人群的身体健康，以及北极、非洲、小岛屿、亚洲及非洲的大三角洲地区等产生影响。

同时政府间气候变化专门委员会指出，人为导致的气候变化可能导致一些突变的或者不可逆的影响。如果全球平均温度增幅超过1.5～2.5℃，则20%～30%的物种面临的灭绝风险将增大；如果全球平均温度升高超过3.5℃，则全球将有40%～70%的物种灭绝。

---

1 ppb：parts per billion，十亿分率。

2 IPCC. 气候变化2007综合报告［M］.瑞典：政府间气候变化专门委员会出版，2008：2–5.

3 IPCC. 气候变化2007综合报告［M］.瑞典：政府间气候变化专门委员会出版，2008：7.

4 IPCC. 气候变化2007综合报告［M］.瑞典：政府间气候变化专门委员会出版，2008：8–13.

由此可见，气候变化对全球产生的不利影响极为严峻。

1.2　《联合国气候变化框架公约》

1.2.1　《联合国气候变化框架公约》的产生

虽然早在19世纪末，瑞典科学家斯万特·阿尔赫尼斯（Svante Arrhenius）便提出了温室效应概念并作出描述，此后许多科学家也逐步认识到由人类活动造成大量温室气体排放所带来的问题，但直到20世纪70年代初期，各国科学家仍很少对气候变化问题进行系统研究。1972年召开的斯德哥尔摩人类环境会议，促进了人们对潜在的气候变化和相关问题的研究。70年代末，科学家们开始把气候变化看做一个潜在的严重问题。1988年世界气象组织和联合国环境规划署共同建立了政府间气候变化专门委员会，同年召开的多伦多会议标志着有关气候变化问题高级辩论的开始，同年12月联合国大会通过了一项关于为人类现在的一代和子孙后代保护气候的决议。1990年6月欧共体的代表在筹备“第二次世界气候大会部长级会议”时，首次提出了保护大气层和控制二氧化碳的主张，随后在同年9月讨论会议的《部长宣言》稿时，又提出了应立即开始“气候变化公约”谈判的主张，并将这一问题纳入《部长宣言》。1992年6月在巴西里约热内卢召开的联合国环境与发展大会上，各国政府签署了《联合国气候变化框架公约》，正式拉开了国际社会共同应对气候变化的大幕[1]。

1.2.2　《联合国气候变化框架公约》的主要内容

《联合国气候变化框架公约》是首个全球共同应对气候变化的框架性文件。它明确提出了要“将温室气体的浓度稳定在防止气候系统受到危险的人为干扰水平上”的温室气体控制目标。同时考虑到“历史上和目前全球温室气体排放的最大部分源自发达国家；发展中国家的人均排放仍相对较低；发展中国家在全球排放中所占的份额将会增加，以满足其社会和发展需

1　庄贵阳，陈迎. 国际气候制度与中国［M］. 北京：世界知识出版社，2005：33-35.

要”，提出了“气候变化的全球性要求所有国家根据其共同但有区别的责任和各自的能力及社会和经济条件，尽可能开展最广泛的合作，并参与有效和适当的国际应对行动”。

《联合国气候变化框架公约》同时明确规定，发达国家要率先采取减排行动，使温室气体排放水平降低到1990年的排放水平，并且向发展中国家提供技术和资金援助,以支持发展中国家开展应对气候变化活动。这种资金和技术援助，应有别于官方发展援助和商业技术转让。发展中国家的义务是编制国家信息通报，核心内容是编制温室气体排放源和吸收汇的国家清单，制定并执行减缓和适应气候变化的国家方案。发展中国家履行上述义务的程度取决于发达国家的资金和技术转让的程度[1]。

《联合国气候变化框架公约》是迄今为止在国际环境与发展领域中影响最大、涉及面最广、意义最为深远的国际法律文件，它涉及人类社会的生产、消费和生活方式，涉及各国国民经济和社会发展的方方面面。但因《联合国气候变化框架公约》仅是定性描述，未能就减排温室气体问题做出具体规定，因此该公约缺乏可操作性。

1 庄贵阳，陈迎. 国际气候制度与中国[M]. 北京：世界知识出版社，2005：41–42.

第 2 章

《京都议定书》与清洁发展机制

为了加强《联合国气候变化框架公约》的可操作性，确保公约内容的全面落实，联合国气候变化框架公约缔约方大会启动了新的进程。

2.1 《京都议定书》

1995年，在柏林举行的第一次联合国气候变化框架公约缔约方大会上通过了“柏林授权”（Berlin Mandate）：认为《联合国气候变化框架公约》对发达国家缔约方和其他缔约方的具体承诺不足，一致同意开始一个新的进程，以使其能够为2000年以后的阶段采取适当行动，包括通过一项决议书或另外一种法律文件，以加强附件一缔约方在《联合国气候变化框架公约》下的承诺效力，并对非附件一缔约方不引入任何新的承诺[1]。

经过历时三年的艰苦谈判，1997年12月1日至11日，联合国气候变化框架公约第三次缔约方大会在日本京都举行。此次会议的宗旨是确定一个具有法律约束力的温室气体减排目标和期限，以使发达国家更有效地降低温室气体的排放量，尽快抑制全球变暖的趋势。本次会议将削减发达国家温室气体的排放量作为核心议题予以审议。会议各方经过激烈和艰难的谈判磋商，通过了具有里程碑意义的《京都议定书》。这使得《联合国气候变化框架公约》的实施向前迈出了极其重要的一步，标志着气候变化国际谈判进入了建设性的发展阶段。同时，《京都议定书》也成为首个为发达国家规定量化

1 联合国气候变化框架公约，柏林授权，1995.

减排义务（包括减排量和减排时限）的法律文件，为全面落实《联合国气候变化框架公约》内容奠定了基础。

2.1.1 《京都议定书》的主要内容

《京都议定书》明确规定，在《联合国气候变化框架公约》及“柏林授权”的相关原则下，以二氧化碳（CO_2）、甲烷（CH_4）、氧化亚氮（N_2O）、氢氟碳化物（HFCs）、全氟碳化物（PFCs）和六氟化硫（SF_6）六种温室气体为控制目标[1]，规定附件一缔约方在2008～2012年承诺期内，上述六种温室气体排放总量从1990年水平至少减少5%。

同时，为降低发达国家减排成本，确保减排目标的实现，《京都议定书》在要求发达国家主要依靠自身减排的同时，规定了联合履约（Joint Implementation，JI）、清洁发展机制（Clean Development Mechanism，CDM）和排放贸易（Emissions Trading，ET）三种灵活的市场履约机制。

《京都议定书》还规定，《京都议定书》应在不少于55个缔约方，且其合计的二氧化碳排放量至少占附件一缔约方1990年二氧化碳排放总量的55%的附件一缔约方已经交存其批准、接受、核准或加入的文书之日后第90天起生效[2]。

2.1.2 《京都议定书》的发展历程

虽然1997年《京都议定书》即获得了通过，但由于《京都议定书》只涉及减排目标和机制，具体实施的技术细节并未涉及，因此随后围绕具体实施和技术细节又展开多轮艰难谈判，直到2004年11月俄罗斯正式签署《京都议定书》后，才使得签署《京都议定书》的附件一缔约方二氧化碳排放量超过附件一缔约方的55%，从而使《京都议定书》满足生效条件。2005年2月16日《京都议定书》正式生效。截至2010年年底，全球共有192个缔约方签署了《京都议定书》，附件一缔约方的二氧化碳排放量占附件一缔约方1990年

1 六种温室气体的全球升温潜势（GWP）分别为：CO_2, 1; CH_4, 21; N_2O, 310; HFCs, 140～11,700; PFCs, 6,500～9,200; SF_6, 23,900.

2 联合国. 联合国气候变化框架公约, 京都议定书, 1998.

二氧化碳排放总量的63.7%[1]。

第四次缔约方大会于1998年11月2日至13日在阿根廷首都布宜诺斯艾利斯举行。会议达成了《布宜诺斯艾利斯行动计划》（Buenos Aires Plan of Action）。该计划维护了“共同但有区别的责任原则”，并要求在第六次缔约方会议上就《京都议定书》的具体实施规则达成协议，争取就推进实施《联合国气候变化框架公约》中有关发达国家向发展中国家提供资金技术援助的承诺达成协议，从而为2002年《京都议定书》的生效做好准备。该计划为《京都议定书》的深入谈判指明了方向。

第五次缔约方会议于1999年10月25日至11月5日在德国波恩举行。会议就《京都议定书》生效所需的具体细则进行磋商，包括发展中国家参与、京都机制、履约程序、碳汇等议题。

第六次缔约方会议于2000年11月13日至24日在荷兰海牙召开。会议的主要目的是确定落实《京都议定书》的具体措施，以切实履行发达国家在《京都议定书》中做出的减排承诺。由于各国利益的相互牵制，会议进展缓慢，尤其是欧盟和以美国为首的伞形国家集团之间在碳贸易、碳汇以及《京都议定书》履约机制上的立场相距甚远，发达国家与发展中国家在技术开发与转让、能力建设和资金机制等问题上存在对立，发展中国家内部在某些问题上也存在分歧。会议最终无法在一系列问题上达成一致意见。于是，会议决定在2001年召开第六次缔约方会议的续会，继续就履约相关内容进行谈判磋商。

第六次缔约方会议续会于2001年7月16日至27日在德国波恩召开，会议达成了《波恩协议》（Bonn Agreement）。会议之前，政府间气候变化专门委员会于2001年1月正式发布第三次评估报告，美国总统布什于2001年3月突然单方面宣布退出《京都议定书》。这些事件既让人们认识到气候变化问题的重要性和复杂性，又让人们体会到气候变化谈判的严峻性。经过艰苦谈判以及各方妥协，续会就《联合国气候变化框架公约》和《京都议定书》下的资金机制、技术开发和转让、京都机制的参与资格、适用范围、执行机构、履约机制，以及土地利用、土地利用变化和林业活动等具体问题，初步达成共识，最终通过了《波恩协议》。《波恩协议》是妥协的产物，发展中国家做出了重大的让步，一些发达国家减轻了自己控制温室气体的责任和向发展

1 联合国气候变化框架公约. Status of Ratification of the Kyoto Protocol. http://unfccc.int/essential_background/kyoto_protocol/status_of_ratification/items/5524.php.

中国家提供经济技术援助的责任。但它使得《京都议定书》免遭夭折，为推动《京都议定书》的生效奠定了政治基础。参与各方在没有美国参与的情况下，本着“共同但有区别的责任原则”，达成了《京都议定书》下一系列具体实施规则，在构建国际气候制度的进程中取得了突破性的进展。

第七次缔约方会议于2001年10月29日至11月10日在摩洛哥马拉喀什举行。会议经过艰苦谈判，以一揽子方式通过了落实《波恩协议》的一系列决议，统称为《马拉喀什协定》（Marrakech Accords）。会议通过了波恩会议就资金机制、技术转让、能力建设等问题形成的决议草案，对第六次缔约方会议续会遗留下来的《京都议定书》三大机制、履约程序和碳汇问题，达成一揽子解决方案，并且在发达国家向发展中国家提供资金援助方面取得较大进展。此次会议使国际气候谈判进入到各缔约方批准《京都议定书》的关键阶段。《马拉喀什协定》体现了从理论到实际的转变，使《京都议定书》由国际谈判真正走向国际行动。

第八次缔约方会议于2002年10月23日至11月1日在印度首都新德里举行。“在可持续发展的框架下应对气候变化问题”是本次会议的核心议题，并提出“针对气候变化的适应性措施”是所有国家在气候变化方面的优先工作。会议通过了《德里宣言》（Delhi Declaration）。按照“共同但有区别的责任原则”，在可持续发展战略下考虑气候变化问题成为世界各国构思其气候变化战略的重要思路。

第九次缔约方会议于2003年12月1日至12日在意大利米兰举行。欧盟在会议上努力游说，试图推动《京都议定书》早日生效，坚持将全球承诺和减排作为其谈判的核心政策，促使发展中国家承担实质性的减排义务。但俄罗斯在本次会议上明确表态，暂时不批准《京都议定书》，使得试图通过拖延《京都议定书》生效时间进一步拖延。

第十次缔约方大会于2004年12月6日至17日在阿根廷首都布宜诺斯艾利斯举行。此次会议正值《联合国气候变化框架公约》生效十周年。通过国际社会的共同努力，俄罗斯于会前一个月前正式签署了《京都议定书》，使这次大会成为《京都议定书》正式生效前最后一次缔约方会议。大会通过了一系列旨在帮助各国为适应气候变化做准备的措施。考虑到已经探测到的气候变化的影响越来越明显，各国都批准了《布宜诺斯艾利斯适应工作计划》。该计划包括有关气候变化风险和适应的诸多研讨会和技术文件，以及支持让适应成为可持续发展规划的主要内容等重要事项。对最不发达国家的“国家适应行动计划”给予支持。会议要求各国政府采取政策和措施支持本国履行对《联合国气候变化框架公约》和《京都议定书》的现有承诺[1]。

2.2 清洁发展机制

2.2.1 清洁发展机制的基本概念

清洁发展机制是《京都议定书》规定的三种灵活减排机制之一。它具有双重目的：帮助未列入附件一的缔约方（发展中国家）实现可持续发展和有益于《联合国气候变化框架公约》的最终目标，同时协助附件一所列缔约方（发达国家）实现遵守《京都议定书》第 3 条规定的其量化的限制和减少排放的承诺[2]。

清洁发展机制的核心是指《京都议定书》发达国家缔约方以通过提供资金和技术的方式，与发展中国家开展项目级的减排合作，而项目所实现的额外的“经核证的减排量”（Certified Emission Reducitons，CERs）可以用于发达国家缔约方完成《京都议定书》减排目标的承诺。

按照《京都议定书》规定，清洁发展机制在2000年以后就可以开始实施，累积的核证减排量可用于附件一国家完成其2008 ~ 2012年第一承诺期的部分减排义务；发展中国家有权依据本国实施可持续发展战略的需求，自行确定清洁发展机制项目的优先领域。双方可在自愿基础上参与清洁发展机制项目，但必须经双边政府批准。

清洁发展机制被普遍认为是一种双赢机制。从理论上看，发达国家通过这种合作，可以以远低于其国内减排所需的成本实现其在《京都议定书》下的减排义务，节约大量的资金，并且可以通过这种方式将技术、产品甚至清洁发展理念输入到发展中国家；发展中国家通过这种项目级的合作，可以获得技术、理念，以及实现减排所需的资金甚至更多的投资，从而促进其经济发展和环境保护，实现其可持续发展的目标。

适合于清洁发展机制项目的技术和措施很多。从广泛意义来看，任何有

---

1 国家发展改革委应对气候变化司. 清洁发展机制读本［M］. 北京：中国标准出版社，2008：4–6.

2 联合国. 联合国气候变化框架公约，京都议定书，1998.

益于温室气体减排和温室气体回收或吸收的技术和措施，都可以申请清洁发展机制项目。根据联合国清洁发展机制执行理事会（Executive Board，EB）有关规定，清洁发展机制项目活动主要划分为如下15个领域[1]：

（1）能源行业（可再生能源和非可再生能源的生产）；

（2）能源分配；

（3）能源需求；

（4）制造业；

（5）化工业；

（6）建筑；

（7）交通运输；

（8）采矿/矿产品；

（9）冶金产品；

（10）燃料挥发排放物（固体、液体和气体）；

（11）氢氟碳化物和六氟化硫产品生产或消耗过程中的挥发物；

（12）溶剂使用；

（13）废物处理处置；

（14）造林和再造林；

（15）农业。

2.2.2 清洁发展机制的国际规则和要求

《京都议定书》第12条对清洁发展机制项目只做了非常原则性的规定。从1998年开始，联合国气候变化框架公约缔约方会议、公约附属科技咨询机构（SBSTA）和附属执行机构（SBI）就制定详细的清洁发展机制项目合作的规则，经过长达4年的艰苦谈判，在2001年的第七次缔约方大会上，终于就实施规则达成一致。这些规则主要体现在第七次缔约方会议第15号和第17号决定及其附件中。随后，清洁发展机制执行理事会依据所赋予的职责，就清洁发展机制实施的大量具体技术性问题做出了一系列决定[2]。

---

1 联合国清洁发展机制执行理事会. http://cdm.unfccc.int/Statistics/Registration/RegisteredProjByScopePieChart.html.

2 国家发展改革委应对气候变化司. 清洁发展机制读本［M］. 北京：中国标准出版社，2008：6-8.

2.2.2.1 参与资格

第七次缔约方会议第17号文件《清洁发展机制的方式和程序》第28～32段对缔约方参加清洁发展机制合作的资格做了详细规定。无论是发达国家还是发展中国家，必须是《京都议定书》的缔约方才能够参与清洁发展机制项目合作。参与清洁发展机制合作必须基于自愿，必须有负责清洁发展机制的国家主管机构。

我国于2002年由国务院核准《京都议定书》。《京都议定书》生效后，在法律资格上，我国成为清洁发展机制的参与方之一。

2.2.2.2 项目合格性标准

《清洁发展机制的方式和程序》第37～52段对清洁发展机制项目的合格性提出了具体要求，主要包括：项目相对于基准线而言必须能够产生温室气体减排量；项目须经参与项目的缔约方政府批准；项目所采用的方法学应是经过批准的方法学；项目如果带来其他环境问题，应提出解决这些环境问题的办法；项目基准线应以项目为基础，并考虑保守和透明的方式来确定；建立基准线还应该充分考虑国家和行业的政策和规则；应该为项目选择合理的边界，并充分考虑项目可能产生的温室气体“泄漏”问题，等等。

我国重点关注以下三项要素：

（1）项目的基准线及产生的经核证的减排量。项目一定要能够产生和带来温室气体减排量。

（2）项目应该带来先进的技术转让。这种技术可以是我国没有的，也可以是我国有但是商业化程度较低或难以商业化的。

（3）项目能够带来资金。这种资金应该是发达国家提供的，额外于发达国家官方发展援助资金。

2.2.2.3 参与机构

清洁发展机制项目执行过程中，主要参与机构有：项目业主、项目所在国政府、核查/核证项目的指定经营实体、清洁发展机制执行理事会以及缔约方会议。

（1）项目业主，负责按照清洁发展机制执行理事会颁布的模板，编制项目设计文件，将文件提交给项目所在国政府批准，并邀请一个获得授权的指定经营实体审定项目。在项目通过审定并获得注册后，执行项目，根据项目设计文件所提出的监测方案监测项目实施情况。在项目执行一段时期后，按要求邀请另一家指定经营实体对项目所产生的温室气体减排量进行核证。

（2）项目所在国政府，负责审查报批的清洁发展机制项目是否符合国家的可持续发展需求和相关政策要求，决定是否批准所申报的将在其境内实施的项目作为清洁发展机制项目。项目所在国政府可以通过颁布政策、建立专门机构等方式，管理其国内机构与其他发达国家机构之间开展的清洁发展机制项目。

（3）指定经营实体，主要职责是依据清洁发展机制的各项规则要求，对项目业主所申请的、作为清洁发展机制的项目进行审定，并在认为合格后提交清洁发展机制执行理事会申请注册；负责在项目执行之后对项目所产生的温室气体减排量进行核证、向清洁发展机制执行理事会申请签发核证减排量。

（4）清洁发展机制执行理事会，监管清洁发展机制项目的实施，主要包括：根据缔约方会议的决定和指导意见，制定具体的清洁发展机制实施细则；提出和批准清洁发展机制项目方法学；委任经营实体并报缔约方会议批准；审批清洁发展机制项目注册申请；签发项目所产生的温室气体核证减排量。

（5）缔约方会议，是清洁发展机制的最高决策机构，也是《联合国气候变化框架公约》及《京都议定书》下所有问题的最高决策机构。缔约方会议由所有缔约方代表参加，每年召开一次，磋商和解决有关气候变化的问题。

2.2.2.4 项目开发流程

一个典型的清洁发展机制项目开发流程如图2-1所示[1]。

1 国家发展改革委应对气候变化司. 清洁发展机制读本［M］. 北京：中国标准出版社，2008：32-34.

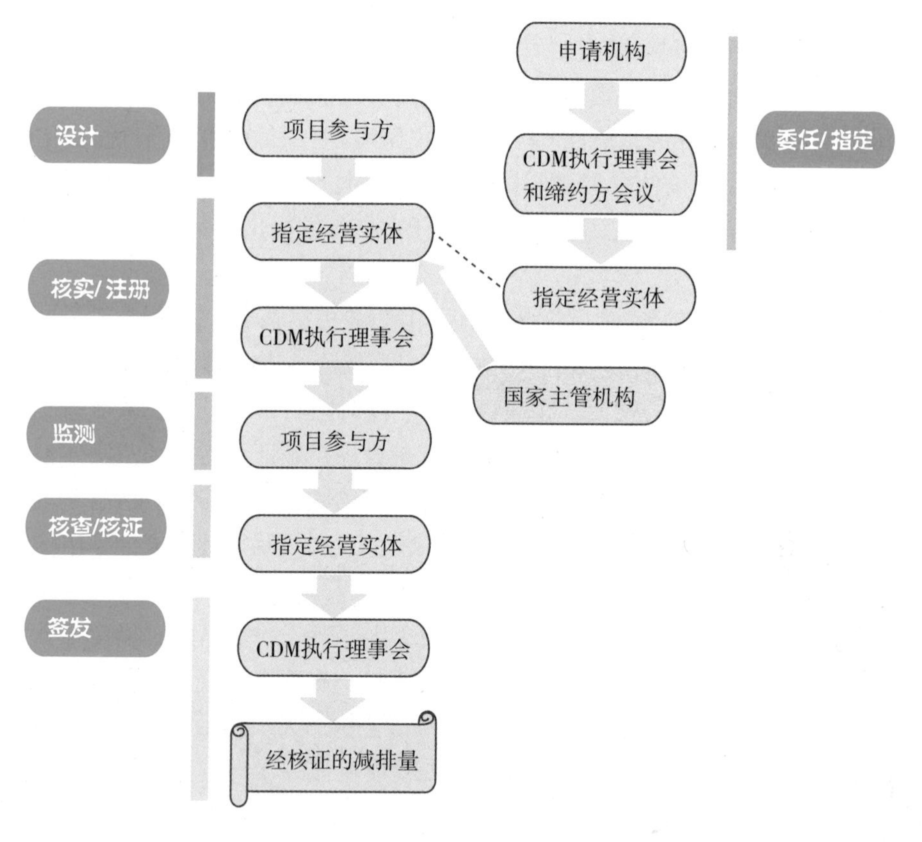

图 2–1 典型清洁发展机制项目开发流程

（1）项目准备。首先，项目业主按照清洁发展机制执行理事会颁布的标准模板，编写项目设计文件。然后，将准备好的项目设计文件提交其政府批准。鉴于该项目设计文件的编写涉及极为专业的知识和能力，且工作语言为英语，目前，在我国项目业主一般会委托专业咨询机构帮助准备项目设计文件。也有个别能力较强的项目业主自行组织项目设计文件的编写。

（2）政府批准。政府主管部门在收到项目设计文件后，组织专家评审，对符合条件和项目设计文件合格的项目出具政府批准函。

（3）项目合格性审查。在政府批准后，项目业主选择一家联合国清洁发展机制执行理事会授权的指定经营实体，邀请其对项目合格性进行审定。项目业主与指定经营实体间签订服务合同，并向指定经营实体提供所需证明材料。受邀请的指定经营实体依据清洁发展机制的各项规则，对所申报项目的合格性进行逐条审定；指定经营实体在确认所申报项目合格后，将项目审定报告提交给清洁发展机制执行理事会，申请该项目正式注册。

（4）项目注册。联合国气候变化框架公约秘书处审核指定经营实体提交的申请注册项目文件的完整性。确认信息完整后，将项目在联合国气候变化框架公约官方网站的清洁发展机制栏目下公示8周。公示期内，如果没有清洁发展机制执行理事会3名及以上成员或项目任一参与方提出质疑，项目可正式获得注册。

（5）项目活动监测。项目获得注册后，项目业主实施项目，并根据项目设计文件所述的监测方案，监测项目实施情况，特别是温室气体减排情况。

（6）减排量核查/核证。在项目执行一段时间后，项目业主邀请另一家指定经营实体对项目所产生的温室气体减排量进行核证。指定经营实体根据项目监测报告，计算出项目实际产生的温室气体减排量，编写包括温室气体减排量在内的项目核证报告，提交联合国清洁发展机制执行理事会，申请签发核证减排量。

（7）核证减排量签发。联合国气候变化框架公约秘书处对指定经营实体提交的核证减排量签发申请进行完整性审核，在联合国气候变化框架公约官方网站的清洁发展机制栏目下将签发申请公示15天。公示期内，如果没有清洁发展机制执行理事会3名及以上成员或项目任一参与方提出质疑，核证减排量将顺利获得签发。

（8）核证减排量转移。核证减排量签发后，暂存入项目东道国国家账户，待买卖双方确认后，扣除捐赠给联合国气候变化适应基金的2%核证减排量后，其余核证减排量转入买家指定账户。

2.3 我国清洁发展机制项目管理

2.3.1 我国应对气候变化的行动和管理框架

我国作为负责任的发展中大国，高度重视应对气候变化工作。一方面在国际应对气候变化谈判中，在全面坚持《联合国气候变化框架公约》和《京都议定书》原则下，发挥积极、建设性作用，推动国际谈判进展；另一方面切实发挥大国作用，全面贯彻落实我国对国际社会的各项承诺。

我国政府早在1992年6月11日，在里约热内卢召开的联合国环境与发展大会上就签署了《联合国气候变化框架公约》，并于1992年11月17日正式批准了该公约，成为最早批准加入该公约的首批缔约方之一。我国政府又在1998年5月29日签署、2002年8月30日核准了《京都议定书》，成为《京都议定书》第37个签约国[1],并全面贯彻落实议定书有关责任。

2007年6月12日，国务院成立由国务院总理温家宝任组长，20余个部委领导任成员的国家应对气候变化领导小组，作为国家应对气候变化的议事协调机构，主要负责研究制定国家应对气候变化的重大战略、方针和对策，统一部署应对气候变化工作，研究审议国际合作和谈判对策，协调解决应对气候变化工作中的重大问题，组织贯彻落实国务院有关节能减排工作的方针政策，统一部署节能减排工作，研究审议重大政策建议，协调解决工作中的重大问题[2]。

同时，我国制定了《中国应对气候变化国家方案》、《中国应对气候变化政策与行动》，设定了节能减排和应对气候变化目标，并将应对气候变化纳入国家发展规划。

2.3.2 我国清洁发展机制项目管理体系

清洁发展机制作为我国有效利用国际资源推进应对气候变化工作的重要手段之一，自实施伊始即受到我国政府的高度重视。

2004年6月30日，在我国清洁发展机制项目实施起步阶段，国家即颁布了《清洁发展机制项目运行管理暂行办法》，并经一年的试行后，于2005年10月12日正式颁布《清洁发展机制项目运行管理办法》，规定了我国清洁发展机制项目申请的许可条件、国家对项目的审批和管理、项目实施程序等，规范项目申请、审批和后期实施流程，保证了项目实施的规范性和有序性。2011年8月3日，根据我国清洁发展机制项目的实施情况，国家发展和改革委员会、科学技术部、外交部和财政部对《清洁发展机制项目运行管理办法》进行了修订，出台了《清洁发展机制项目运行管理办法（修订）》，以进一

1 UNFCCC. Status of Ratification of the Kyoto Protocol. http://unfccc.int/essential_background/kyoto_protocol/status_of_ratification/items/5524.php.

2 国务院. 国务院关于成立国家应对气候变化及节能减排工作领导小组的通知. 2007年6月12日.

步推进清洁发展机制项目在我国的有序、健康发展。

同时，为加强国家对清洁发展机制项目的管理，确保项目质量，维护国家利益及我国项目的国际声誉，我国政府在国家气候变化对策协调小组下设立由国家发展和改革委员会及科学技术部为组长单位，外交部为副组长单位，环境保护部、中国气象局、财政部和农业部为成员单位的国家清洁发展机制项目审核理事会，具体负责：对申报的清洁发展机制项目进行审核，提出审核意见；向国家气候变化对策协调小组报告清洁发展机制项目执行情况和实施过程中的问题和建议，提出涉及国家清洁发展机制项目运行规则的建议。国家发展和改革委员会作为我国开展清洁发展机制项目的主管机构，具体负责：组织受理清洁发展机制项目申请；依据项目审核理事会的审核意见，会同科学技术部和外交部批准清洁发展机制项目；出具清洁发展机制项目批准文件；组织对清洁发展机制项目实施监督管理等相关事务。我国清洁发展机制项目国家管理架构如图2–2所示。

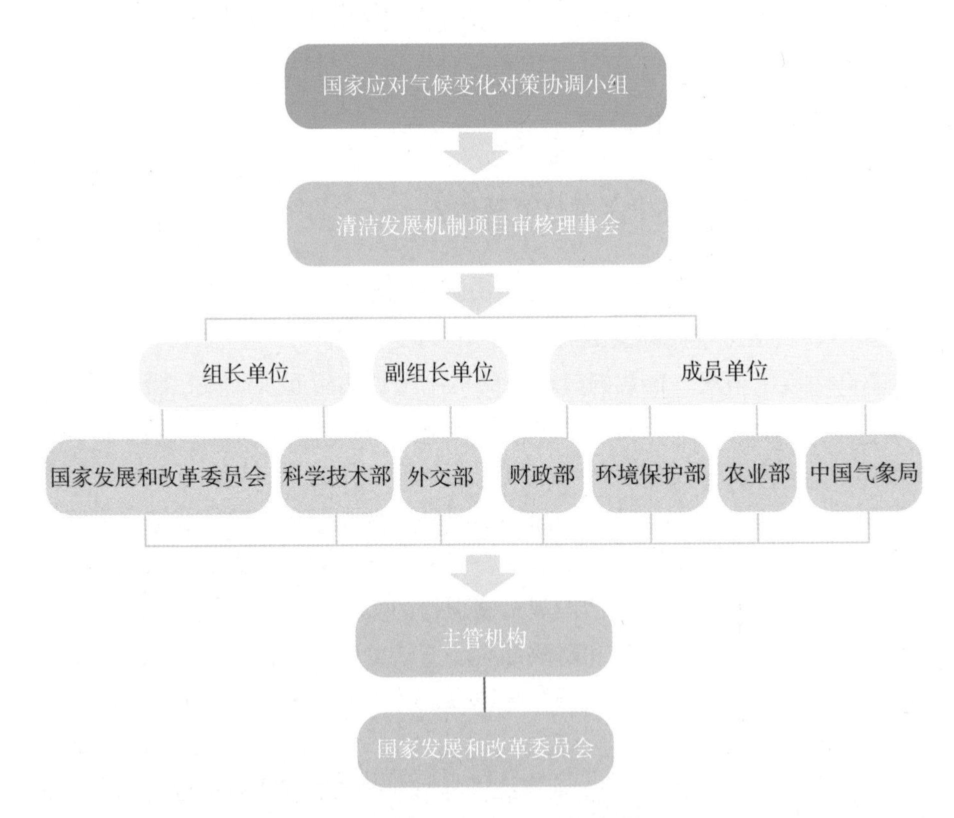

图 2–2　我国清洁发展机制项目管理架构

第3章

清洁发展机制发展状况

清洁发展机制作为三种灵活履约机制中唯一联系发达国家与发展中国家的减排机制，自2004年实施以来，历经数年发展，已成为当前国际上发达国家和发展中国家共同应对气候变化发展最成熟机制。

3.1 全球清洁发展机制项目发展状况

根据清洁发展机制有关规则，清洁发展机制项目的开发和实施先后须经历国家批准、联合国清洁发展机制执行理事会批准注册，以及联合国清洁发展机制执行理事会签发项目产生的核证减排量三个关键环节，才能最终获得核证减排量的交易。统计、分析这三个关键环节的进展状况，有助于全面了解清洁发展机制项目的总体进展状况。由于国家批准阶段属各国内部事务，在公开渠道难以获得各国国家批准情况的详尽数据，故对全球清洁发展机制项目的发展状况分析，着重于项目注册和核证减排量签发两个环节。

3.1.1 项目注册情况

3.1.1.1 注册现状

截至2011年6月30日，全球共有3,368个清洁发展机制项目获得联合国清洁发展机制执行理事会批准，注册成功。这些项目分布在71个发展中国家。

项目顺利实施后，预计每年可产生的温室气体减排量（预期年减排量）达5.07亿吨二氧化碳当量（tCO_2e）。

按项目数统计，这些项目主要分布在中国、印度、巴西、墨西哥、马来西亚和印度尼西亚等国家。上述6个国家的项目数占全球注册项目总数的81.2%。其中，中国以占据全球项目数45.0%的优势遥遥领先。全球注册项目数的分布状况详见表3-1。

表 3-1　　全球注册项目数分布状况

国　家	项目数（个）	比例（%）
中　国	1,516	45.0
印　度	709	21.0
巴　西	212	6.3
墨西哥	130	3.9
马来西亚	97	2.9
印度尼西亚	70	2.1
其　他	634	18.8
合　计	3,368	100.0

资料来源：根据联合国气候变化框架公约清洁发展机制执行理事会网站数据整理而得。

按项目预期年减排量统计，这些项目主要分布在中国、印度、巴西、韩国、墨西哥和印度尼西亚等国家。上述6个国家的项目预期年减排量占注册项目预期年减排总量的86.3%。其中，中国以占全球预期年减排总量的63.5%居于首位。全球注册项目的预期年减排量分布状况详见表3-2。

表 3-2　　全球注册项目预期年减排量分布状况

国　家	预期年减排量（$MtCO_2e$）	比例（%）
中　国	318.67	63.5
印　度	53.02	10.6
巴　西	24.31	4.8
韩　国	18.72	3.7
墨西哥	10.49	2.1

续表

国　家	预期年减排量（$MtCO_2e$）	比例（%）
印度尼西亚	7.53	1.5
其　他	68.70	13.7
合　计	507.23	100.0

注：$1MtCO_2e$=1,000,000tCO_2e。

资料来源：同表3-1。

3.1.1.2　注册项目发展趋势

自2004年11月18日全球首个项目注册成功后，全球清洁发展机制项目快速发展。2008年2月26日，全球注册项目数首次突破1,000个，历时1,195天。2009年12月13日突破2,000个，历时656天。2011年2月2日突破3,000个，历时仅416天。截至2011年6月30日，注册项目数达3,369个。依此速度，注册项目数突破4,000个的用时预计为400天左右。注册项目数每突破1,000个的用时情况详见图3-1。

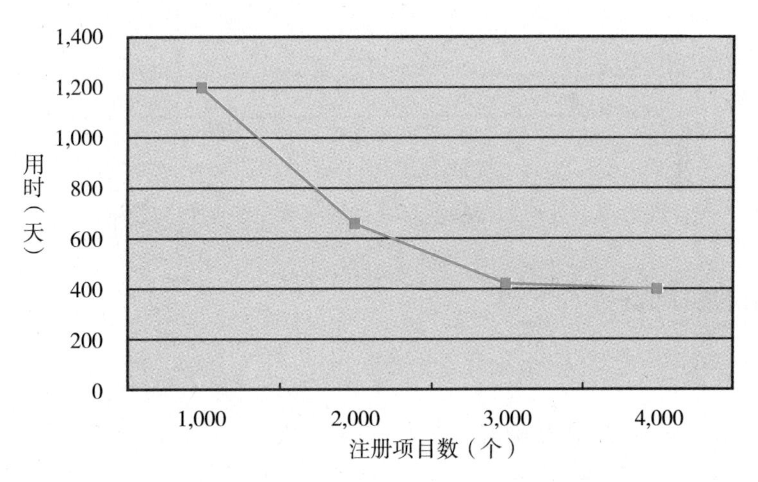

图 3-1　注册项目突破千个的用时情况

资料来源：根据联合国气候变化框架公约清洁发展机制执行理事会网站数据整理而得。

按预期年减排量统计，自2004年11月18日全球首个清洁发展机制项目注册成功，到2006年10月27日，全球注册项目预期年减排量首次突破1亿吨二氧化碳当量，历时708天。2008年2月1日突破2亿吨二氧化碳当量，历时462天。2009年4月20日突破3亿吨二氧化碳当量，历时444天。2010年10月16日突破4亿吨二氧化碳当量，历时544天。2011年4月28日突破5亿吨二氧化碳当量，历时仅194天。注册项目的预期年减排量突破4亿吨的用时较之前用时增加的主要原因是，全球HFC-23分解等预期年减排量较大的工业废气类项目已基本开发完毕，而其余项目规模普遍较小，联合国清洁发展机制执行理事会及指定经营实体的工作效率又无显著提高。注册项目预期年减排量突破5亿吨的用时大幅缩减，则是联合国清洁发展机制执行理事会大幅提升工作效率的结果。依目前的注册情况推断，突破6亿吨的用时将继续缩减。注册项目预期年减排量突破亿吨的用时情况详见图3-2。

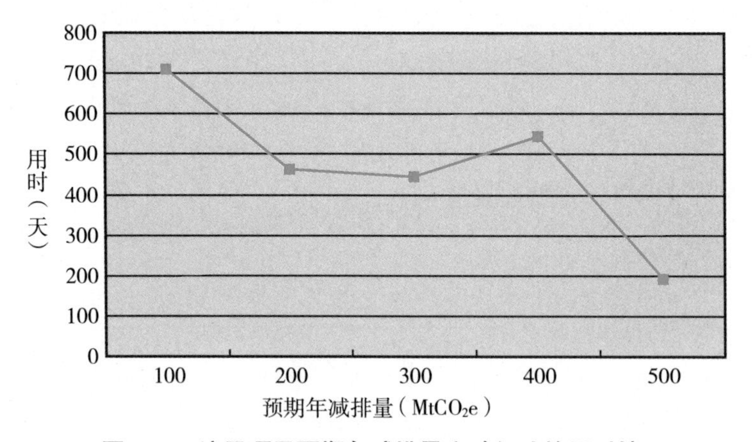

图 3-2 注册项目预期年减排量突破亿吨的用时情况

资料来源：同图3-1。

3.1.2 核证减排量签发情况

3.1.2.1 核证减排量签发现状

截至2011年6月30日，全球共有2,789批次6.47亿吨二氧化碳当量的核证减排量获得签发。以核证减排量交易价格平均8美元/吨二氧化碳当量估算，清洁发展机制共为发展中国家带来可持续发展资金约52亿美元[1]。这些核证减排量主要分布在44个发展中国家，其中，中国、印度、韩国、巴西、墨西哥和智利等6个国家的签发量占全球已签发总量的93.9%。中国的核证减排量为3.68亿吨，占全球签发总量的56.9%，居全球首位。全球核证减排量分布状况详见表3–3。

表 3–3　全球核证减排量分布状况

国　家	签发批次		签发量	
	批次	比例（%）	签发量($MtCO_2e$)	比例（%）
中　国	1,125	40.3	368.07	56.9
印　度	672	24.1	97.76	15.1
韩　国	105	3.8	71.79	11.1
巴　西	389	13.9	54.10	8.4
墨西哥	125	4.5	8.34	1.3
智　利	55	2.0	7.08	1.1
其　他	318	11.4	39.59	6.1
合　计	2,789	100.0	646.73	100.0

注：1 $MtCO_2e$=1,000,000tCO_2e。

资料来源：同表3–1。

3.1.2.2　核证减排量签发趋势

自2005年10月20日全球首笔核证减排量（CERs）获得签发，到2007年12月14日，签发的核证减排量首次突破1亿吨二氧化碳当量，历时785天。2008年10月16日签发量突破2亿吨，历时307天。2009年6月23日签发量突破3亿吨，历时250天。2010年4月9日签发量突破4亿吨，历时290天。2011年1月10日签发量突破5亿吨，历时276天。2011年4月27日签发量突破6亿吨，历时107天。截至2011年6月30日，签发量已达6.85亿吨，历时仅64天，预计突破7亿吨的用时将进一步缩短。核证减排量签发每突破亿吨的用时情况

1　在早期签署的清洁发展机制项目减排量交易协议中，核证减排量的交易价格较低，一般在6～8美元/吨二氧化碳当量之间，后期的交易价格均较高。

详见图3-3。从中可以看出，随着项目的逐步开展，核证减排量签发速度逐步加快，签发量突破亿吨的用时总体呈现下降趋势。但突破4亿吨和5亿吨的用时略有上升后，突破6亿吨的用时快速下降。这主要是与2010年下半年联合国清洁发展机制执行理事会暂停所有HFC-23分解类清洁发展机制项目核证减排量签发，又于2010年12月恢复签发，导致2010年下半年签发量较少，而相应减排量签发顺延至2011年上半年所造成。

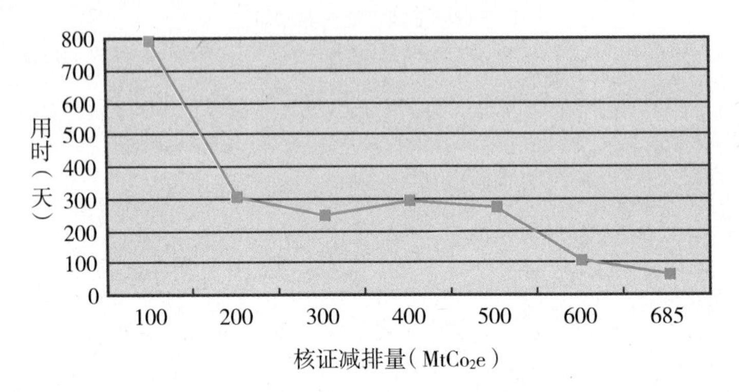

图 3-3 核证减排量签发突破亿吨的用时情况

注：$1MtCO_2e=1,000,000tCO_2e$。
资料来源：同图3-1。

全球清洁发展机制项目的分布情况详见表3-4。

表 3-4 全球清洁发展机制项目分布状况

国家	注册情况				核证减排量签发情况			
	项目数	比例（%）	预期年减排量(tCO_2e)	比例（%）	批次	比例（%）	签发量(tCO_2e)	比例（%）
中国	1,516	45.0	318,665,043	63.5	1125	40.3	368,066,457	56.9
印度	709	21.0	53,023,555	10.5	672	24.1	97,758,431	15.1
巴西	212	6.3	24,312,766	4.8	389	13.9	54,101,186	8.4
墨西哥	130	3.9	10,490,396	2.1	125	4.5	8,340,029	1.3
马来西亚	97	2.9	5,589,842	1.1	22	0.8	1,216,896	0.2
印度尼西亚	70	2.1	7,532,212	1.5	21	0.8	2,604,760	0.4
越南	67	2.0	3,494,109	0.7	5	0.2	6,646,339	1.0

续表

国家	注册情况				核证减排量签发情况			
	项目数	比例（%）	预期年减排量(tCO_2e)	比例（%）	批次	比例（%）	签发量(tCO_2e)	比例（%）
韩　国	60	1.8	18,723,184	3.7	105	3.8	71,790,390	11.1
菲律宾	55	1.6	2,158,700	0.4	7	0.3	240,036	0.0
泰　国	54	1.6	3,139,308	0.6	5	0.2	851,541	0.1
智　利	50	1.5	5,707,230	1.1	55	2.0	7,077,779	1.1
哥伦比亚	32	0.9	3,643,079	0.7	21	0.8	940,317	0.1
秘　鲁	24	0.7	2,757,210	0.5	14	0.5	609,611	0.1
阿根廷	23	0.7	4,917,604	1.0	36	1.3	6,478,615	1.0
以色列	22	0.7	2,223,049	0.4	17	0.6	866,907	0.1
洪都拉斯	21	0.6	371,572	0.1	27	1.0	522,629	0.1
南　非	19	0.6	3,247,426	0.6	13	0.5	1,900,276	0.3
厄瓜多尔	16	0.5	1,371,456	0.3	22	0.8	1,114,540	0.2
巴基斯坦	12	0.4	1,774,587	0.3	11	0.4	2,982,626	0.5
危地马拉	11	0.3	864,760	0.2	15	0.5	947,952	0.1
乌兹别克斯坦	11	0.3	4,402,064	0.9	0	0.0		0.0
埃　及	9	0.3	3,068,050	0.6	16	0.6	6,367,204	1.0
哥斯达黎加	8	0.2	418,606	0.1	6	0.2	320,463	0.0
巴拿马	7	0.2	358,513	0.1	1	0.0	60,180	0.0
塞浦路斯	7	0.2	300,889	0.1	0	0.0		0.0
斯里兰卡	7	0.2	210,168	0.0	8	0.3	237,690	0.0
萨尔迪瓦	6	0.2	619,535	0.1	8	0.3	790,253	0.1
乌拉圭	6	0.2	354,713	0.1	1	0.0	40,613	0.0
伊　朗	6	0.2	692,684	0.1	0	0.0	0	0.0
阿拉伯联合酋长国	5	0.1	356,416	0.1	1	0.0	79,960	0.0
柬埔寨	5	0.1	150,948	0.0	1	0.0	10,758	0.0
肯尼亚	5	0.1	1,201,980	0.2	0	0.0		0.0
摩洛哥	5	0.1	287,447	0.1	4	0.1	330,099	0.1
尼加拉瓜	5	0.1	585,296	0.1	10	0.4	577,757	0.1
尼日利亚	5	0.1	4,693,552	0.9	1	0.0	1,867	0.0
亚美尼亚	5	0.1	223,063	0.0	1	0.0	12,022	0.0

续表

国家	注册情况				核证减排量签发情况			
	项目数	比例（%）	预期年减排量(tCO_2e)	比例（%）	批次	比例（%）	签发量(tCO_2e)	比例（%）
玻利维亚	4	0.1	563,991	0.1	3	0.1	1,117,802	0.2
摩尔多瓦	4	0.1	226,585	0.0	0	0.0	0	0.0
尼泊尔	4	0.1	154,317	0.0	0	0.0	0	0.0
乌干达	4	0.1	117,550	0.0	0	0.0	0	0.0
阿拉伯叙利亚共和国	3	0.1	320,782	0.1	0	0.0	0	0.0
科特迪瓦	3	0.1	639,203	0.1	0	0.0	0	0.0
卢旺达	3	0.1	29,682	0.0	0	0.0	0	0.0
蒙　古	3	0.1	71,904	0.0	1	0.0	48	0.0
巴拉圭	2	0.1	18,711	0.0	0	0.0	0	0.0
不　丹	2	0.1	499,522	0.1	1	0.0	474	0.0
多米尼加共和国	2	0.1	483,726	0.1	1	0.0	11,637	0.0
斐　济	2	0.1	47,399	0.0	2	0.1	35,550	0.0
刚果共和国	2	0.1	179,330	0.0	0	0.0		0.0
格鲁吉亚	2	0.1	411,897	0.1	1	0.0	53,138	0.0
古　巴	2	0.1	465,397	0.1	2	0.1	171,178	0.0
喀麦隆	2	0.1	193,462	0.0	0	0.0	0	0.0
孟加拉国	2	0.1	169,259	0.0	0	0.0	0	0.0
突尼斯	2	0.1	687,573	0.1	0	0.0	0	0.0
新加坡	2	0.1	116,782	0.0	0	0.0	0	0.0
约　旦	2	0.1	434,074	0.1	4	0.1	985,992	0.2
阿尔巴尼亚	1	0.0	22,964	0.0	0	0.0		0.0
埃塞俄比亚	1	0.0	29,343	0.0	0	0.0		0.0
巴布亚新几内亚	1	0.0	278,904	0.1	2	0.1	215,424	0.0
圭亚那	1	0.0	44,733	0.0	0	0.0	0	0.0
卡塔尔	1	0.0	2,499,649	0.5	0	0.0	0	0.0
老　挝	1	0.0	3,338	0.0	1	0.0	2,168	0.0
利比亚	1	0.0	93,635	0.0	0	0.0	0	0.0
马达加斯加	1	0.0	44,196	0.0	0	0.0	0	0.0

续表

国　家	注册情况				核证减排量签发情况			
	项目数	比例（%）	预期年减排量(tCO_2e)	比例（%）	批次	比例（%）	签发量(tCO_2e)	比例（%）
马　里	1	0.0	188,282	0.0	0	0.0	0	0.0
马其顿共和国	1	0.0	54,623	0.0	0	0.0	0	0.0
塞内加尔	1	0.0	37,386	0.0	0	0.0	0	0.0
坦桑尼亚联合共和国	1	0.0	202,271	0.0	2	0.1	35,122	0.0
牙买加	1	0.0	52,540	0.0	4	0.1	211,223	0.0
赞比亚	1	0.0	130,032	0.0	0	0.0		0.0
合　计	3,369	100.0	507,233,736	100.0	2,789	100.0	646,725,939	100.0

注：截至2011年6月30日。

资料来源：同表3-1。

全球清洁发展机制项目发展速度

图3-4是以半年作为时间尺度统计的全球清洁发展机制项目注册和签发情况。从图上可见，自2004年清洁发展机制项目实施开始至2006年底，全球清洁发展机制项目处于快速发展阶段。那时，因项目数量相对较少，联合国清洁发展机制项目执行理事会工作处于尚未饱和状态，故注册项目数处于快速增加阶段。但自2007年下半年起，项目注册速度有所降低，由2006年下半年及2007年上半年的250个左右降至174个，并在此后维持在200～300个。

随着各国对清洁发展机制项目的重视和项目开发熟练程度提高，各国申请注册项目数快速增加，联合国清洁发展机制执行理事会的工作效率开始成为制约项目快速发展的主要因素之一，受到各方诟病。为此，联合国清洁发展机制执行理事会自2010年下半年起逐步优化项目审批流程，加快审批速度，并在2010年底坎昆召开的联合国气候变化大会上通过的《坎昆协议》一揽子协议中，出台了《关于清洁发展机制项目的进一步指南》，进一步明确和优化清洁发展机制项目实施流程，设定项目审批时限，督促清洁发展机制项目实施的各环节提高效率。从2010年下半年（548个）及2011年上半年（517个）注册项目数量看，审批速度显著提升，半年内审批项目由此前的200～300个增至500个以上。

从全球每半年内核证减排量签发情况来看，随着全球清洁发展机制项目的实施，核证减排量签发工作的趋于饱和，核证减排量的签发批次由快速增加转为相对平稳。2010年每半年的签发批次维持在309次。但随着2010年底《关于清洁发展机制项目的进一步指南》的出台，全球核证减排量的签发速度显著提升，2011年上半年达734批次，签发量达1.5亿吨二氧化碳当量，是之前签发量的2～3倍。

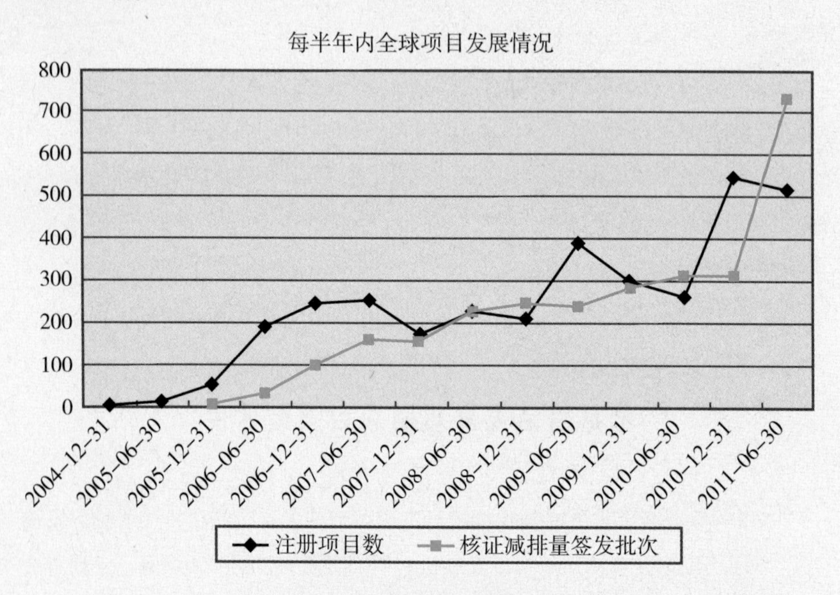

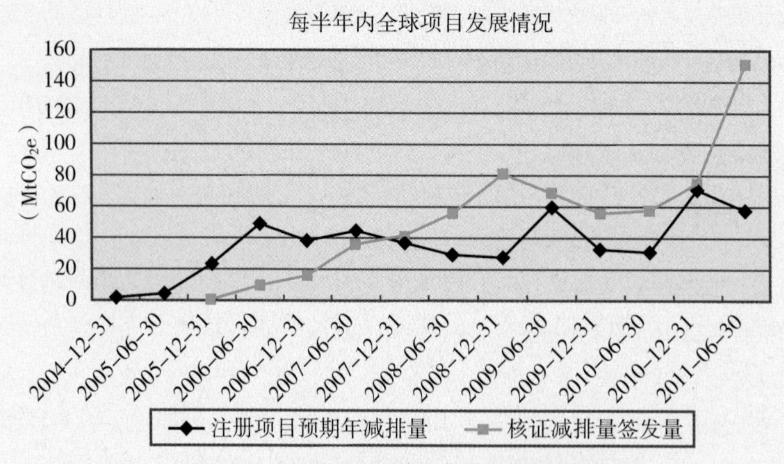

图 3-4　全球清洁发展机制项目发展趋势

资料来源：同图3-1。

3.2　我国清洁发展机制项目发展状况

由于我国政府高度重视推广、利用清洁发展机制，加上我国特定的国情，使得我国成为当前全球清洁发展机制项目发展最为迅速和成熟的国家之一。

3.2.1　我国清洁发展机制项目类型

我国清洁发展机制项目主要集中在以下11类领域:

（1）新能源和可再生能源，主要包括风电、水电、生物质发电、太阳能、地热能、潮汐能等;

（2）节能和提高能效，主要包括节能技改、余热余压利用等;

（3）甲烷回收利用，主要包括煤层气和垃圾填埋气回收利用、户用沼气等;

（4）燃料替代，主要包括天然气代替燃煤发电、锅炉燃煤替代等;

（5）原料替代，主要包括废弃物代替石灰石制水泥等;

（6）垃圾处理，主要包括垃圾焚烧、垃圾堆肥等;

（7）资源回收利用，主要指对工业废气、废料中有价值资源进行回收再利用;

（8）三氟甲烷（HFC-23）分解，主要指将氟化工产品二氟一氯甲烷（HCFC-22）生产过程中产生的高温室效应副产物三氟甲烷，经焚烧或催化处理，转化为无温室效应或者低温室效应气体;

（9）氧化亚氮（N_2O）分解，主要指将硝酸、己二酸等生产过程中产生的高温室效应副产物氧化亚氮，经焚烧或催化处理，转化为无温室效应或者低温室效应气体;

（10）六氟化硫（SF_6）回收利用;

（11）造林再造林。

3.2.2 国家批准项目情况

自2005年1月25日首个清洁发展机制项目获得国家批准至2011年6月30日，我国共有3,104个项目获得国家主管机构批准，项目的预期年减排达4.90亿吨二氧化碳当量。

（1）地域分布状况。这些项目分布在除西藏自治区以外的所有省（市、自治区），未出现一省或几省独大的分布状况。按项目数统计，位居前六位的省份依次是云南、四川、内蒙古、甘肃、湖南和山东。这6个省份的项目数占国家已批准项目总数的43.7%。但由于不同省份资源不同，实施的项目类型不同（如东部地区主要为工业类项目，而西南地区主要为水电、风电等新能源可再生能源类项目），项目产生的减排规模不同，因此，国家批准项目的预期年减排量在各省分布状况与项目数分布状况明显不同。按项目预期年减排量统计，位于前六位的省份依次是四川、江苏、内蒙古、浙江、山西和山东。这6个省份的项目预期年减排量占国家已批准项目预期年减排总量的45.7%。国家已批准项目在各省分布状况详见表3-5和表3-6。

表 3-5　　国家批准项目在各省分布状况

省　份	项目数（个）	比例（%）
云　南	327	10.5
四　川	290	9.3
内蒙古	253	8.2
甘　肃	165	5.3
湖　南	164	5.3
山　东	157	5.1
其　他	1,748	56.3
合　计	3,104	100.0

资料来源：根据中国清洁发展机制网数据整理而得。

表 3-6　　国家批准项目预期年减排量在各省分布状况

省　份	预期年减排量（$MtCO_2e$）	比例（%）
四　川	51.71	9.9
江　苏	41.51	7.9
内蒙古	38.58	7.4
浙　江	37.78	7.2
山　西	35.45	6.8
山　东	34.74	6.6
其　他	284.83	54.3
合　计	524.60	100.0

注：1 $MtCO_2e$=1,000,000tCO_2e。

资料来源：同表3-5。

（2）行业分布状况。按项目数统计，国家批准项目主要集中在新能源和可再生能源、节能和提高能效，以及甲烷回收利用3个行业。这3个行业的项目数占国家批准项目总数的95.6%。按预期年减排量统计，项目主要分布在新能源和可再生能源类、节能和提高能效类、HFC-23分解、甲烷回收利用、N_2O分解以及燃料替代等6个行业。这6个行业的项目预期年减排量占国家批准项目预期年减排总量的98.7%。需要指出的是，虽然HFC-23分解及N_2O分解两类项目的项目数分别仅为11个和32个，但因HFC-23和N_2O的全球升温潜势（Global Warming Potential，GWP）较强（分别为CO_2的11,700倍和310倍），它们单个项目的减排规模巨大，故这两类项目预期年减排量分别排在第3位和第5位。国家批准项目的行业分布状况详见表3-7和表3-8。

表 3-7　　国家批准项目的行业分布状况

行　业	项目数（个）	比例（%）
新能源和可再生能源	2,229	71.8
节能和提高能效	521	16.8
甲烷回收利用	216	7.0

续表

行　业	项目数（个）	比例（%）
燃料替代	42	1.4
原料替代	31	1.0
N_2O分解	32	1.0
HFC-23分解	11	0.4
垃圾处理	9	0.3
资源回收利用	7	0.2
造林和再造林	4	0.1
SF_6回收利用	2	0.1
合　计	3,104	100.0

资料来源：同表3-5。

表 3-8　　国家批准项目预期年减排量的行业分布状况

项目类型	预期年减排量（$MtCO_2e$）	比例（%）
新能源和可再生能源	277.30	52.9
节能和提高能效	83.65	15.9
HFC-23分解	66.90	12.8
甲烷回收利用	52.01	9.9
N_2O分解消除	24.52	4.6
燃料替代	13.53	2.6
原料替代	4.89	0.9
垃圾处理	1.07	0.2
SF_6回收利用	0.32	0.1
资源回收利用	0.30	0.1
造林和再造林	0.12	0.0
合　计	524.60	100.0

注：1 $MtCO_2e$=1,000,000tCO_2e。

资料来源：同表3-5。

3.2.3　项目注册情况

自2005年6月26日我国首个清洁发展机制项目注册成功至2011年6月30日，我国共有1,516个项目获得联合国清洁发展机制项目执行理事会批准注册成功，占联合国已批准注册项目数（3,368个）的45.0%；项目的预期年减排量达3.19亿吨二氧化碳当量，占全球已注册项目预期年减排总量（5.07亿吨二氧化碳当量）的63.5%，高居全球首位。

（1）地域分布状况。按项目数统计，我国注册项目遍布于除西藏自治区以外的所有省（市、自治区），排名前六位的依次为云南、内蒙古、四川、甘肃、湖南和山东，这6个省份的注册项目数占我国注册项目总数的47.8%。按注册项目的预期年减排量统计，排名前六位的依次为浙江、江苏、四川、内蒙古、山东和山西6省份。这一排名与按项目数统计的排名有很大区别。其中，除内蒙古是因项目数多而其注册项目的预期年减排量排名靠前外，其他5省份都是因为有减排规模较大的HFC-23分解及N_2O分解项目，使其注册项目预期年减排量排名靠前。这6个省份的项目预期年减排量占注册项目预期年减排量总量的54.7%。我国注册项目在各省分布状况详见表3-9和表3-10。

表 3-9　我国注册项目在各省分布状况

省　份	项目数（个）	比例（%）
云　南	171	11.3
内蒙古	162	10.7
四　川	139	9.2
甘　肃	106	7.0
湖　南	86	5.7
山　东	60	4.0
河　北	58	3.8
其　他	734	48.4
合　计	1,516	100.0

资料来源：同表3-5。

表 3-10　　我国注册项目预期年减排量在各省分布状况

省　份	预期年减排量（$MtCO_2e$）	比例（%）
浙　江	32.67	10.3
江　苏	31.45	9.9
四　川	27.72	8.7
内蒙古	25.48	8.0
山　东	22.83	7.2
山　西	20.11	6.3
其　他	158.41	49.7
合　计	318.67	100.0

注：1 $MtCO_2e$=1,000,000tCO_2e。

资料来源：同表3-5。

（2）行业分布状况。按项目数统计，我国注册项目主要分布在新能源和可再生能源、节能和提高能效以及甲烷回收利用3个行业。这3个行业项目的数量占我国注册项目总数的94.7%。按项目预期年减排量统计，主要集中在新能源和可再生能源、HFC-23分解、燃料替代、甲烷回收利用、节能和提高能效以及N_2O分解等6个行业。这些项目占注册项目预期年减排总量的99.5%以上。我国注册项目的行业分布状况详见表3-11及表3-12。

表 3-11　　我国注册项目数的行业分布状况

行　业	项目数（个）	比例（%）
新能源和可再生能源	1,220	80.5
节能和提高能效	109	7.2
甲烷回收利用	107	7.1
其　他	80	5.3
合　计	1,516	100.0

资料来源：同表3-5。

表 3–12　我国注册项目预期年减排量的行业分布状况

行　业	预期年减排量（$MtCO_2e$）	比例（%）
新能源和可再生能源	149.12	46.8
HFC–23分解	65.65	20.6
燃料替代	30.03	9.4
甲烷回收利用	28.36	8.9
节能和提高能效	22.76	7.1
N_2O分解	21.02	6.6
其　他	1.72	0.5
合　计	318.67	100.0

注：1 $MtCO_2e$=1,000,000tCO_2e。
资料来源：同表3–5。

3.2.4　核证减排量签发情况

自2006年7月3日我国首笔核证减排量获得签发至2011年6月30日，我国共有489个项目的1,125笔3.68亿吨二氧化碳当量的核证减排量获得签发，占全球已签发核证减排量总量（6.47亿吨二氧化碳当量）的56.9%，远超印度（15.1%），居全球首位。

（1）地域分布状况。除西藏外的各省（市、自治区）均有核证减排量获得签发。排名前六位的省份是浙江、江苏、山东、辽宁、四川和内蒙古，其共同特点是都拥有减排规模大的HFC–23及N_2O分解类项目。这6个省份的核证减排量占我国总量的81.3%。我国核证减排量在各省分布状况详见表3–13。

表 3–13　我国核证减排量在各省分布状况

省　份	项目数（个）	签发批次（笔）	签发量（$MtCO_2e$）	比例（%）
浙　江	16	75	88.93	24.2
江　苏	19	82	88.57	24.1
山　东	18	71	59.35	16.1

续表

省份	项目数（个）	签发批次（笔）	签发量（$MtCO_2e$）	比例（%）
辽宁	17	46	34.39	9.3
四川	37	77	15.62	4.2
内蒙古	50	99	12.29	3.3
其他	332	675	68.91	18.7
合计	489	1,125	368.07	100.0

注：1 $MtCO_2e$=1,000,000tCO_2e。

资料来源：同表3-5。

（2）行业分布状况。我国的核证减排量主要分布在HFC-23分解、新能源和可再生能源以及N_2O分解这3类项目。这3类项目的核证减排量占我国总量的90.7%。其中，HFC-23分解类项目占我国总量的62.6%，远高于其他类型的项目。这主要是因为：第一，HFC-23分解、N_2O分解项目规模较大，实施较早；第二，单个新能源和可再生能源类项目的规模虽不大，但其数量众多。我国已获得的核证减排量的行业分布状况详见表3-14。

表 3-14　我国核证减排量的行业分布状况

行业	项目数（个）	签发批次（笔）	签发量（$MtCO_2e$）	比例（%）
HFC-23分解	11	138	230.41	62.6
新能源和可再生能源	377	738	59.77	16.2
N_2O分解	12	48	43.51	11.8
燃料替代	16	47	14.52	3.9
节能和提高能效	45	90	12.93	3.5
甲烷回收利用	28	64	6.92	1.9
合计	489	1,125	368.07	100.0

注：1 $MtCO_2e$=1,000,000tCO_2e。

资料来源：同表3-5。

我国清洁发展机制项目的实施总体情况详见表3–15和表3–16。

表 3–15　　我国清洁发展机制项目地域分布状况

省份	国家批准		注册		核证减排量签发		
	项目数（个）	预期年减排量($MtCO_2e$)	项目数（个）	预期年减排量($MtCO_2e$)	项目数（个）	批次（笔）	签发量（$MtCO_2e$）
云　南	327	33.5	171	18.6	47	75	4.86
四　川	290	51.7	139	27.7	37	77	15.62
内蒙古	253	38.6	162	25.5	50	99	12.29
甘　肃	165	23.9	106	15.2	25	50	5.24
湖　南	164	15.0	86	8.5	35	64	6.04
山　东	157	34.7	60	22.8	18	71	59.35
河　北	147	19.0	58	8.2	21	59	5.40
山　西	121	35.4	53	20.1	12	36	6.82
河　南	117	19.0	34	9.5	14	35	11.38
浙　江	105	37.8	32	32.7	16	75	88.93
湖　北	99	9.9	50	6.7	13	22	1.66
江　苏	97	41.5	40	31.4	19	82	88.57
辽　宁	97	27.7	48	19.8	17	46	34.39
贵　州	91	9.0	44	2.6	18	29	1.38
吉　林	91	12.9	40	5.7	11	31	2.61
广　东	84	11.3	42	9.6	20	36	2.58
福　建	83	10.6	50	9.3	21	32	2.61
广　西	81	11.7	42	4.5	11	15	0.96
黑龙江	73	15.7	23	4.8	11	25	1.40
陕　西	72	8.5	35	4.4	4	4	0.17
新　疆	69	11.2	43	7.2	12	32	2.72
安　徽	58	8.6	23	3.4	12	21	2.25
重　庆	58	10.2	35	6.5	9	28	1.76
江　西	56	5.0	31	2.3	10	15	0.88
宁　夏	55	5.4	27	3.2	10	30	2.97
青　海	28	2.1	14	1.3	5	13	1.11
上　海	19	6.9	6	3.4	1	1	0.03
海　南	18	0.9	10	0.6	4	8	0.22
北　京	16	4.7	9	2.8	5	13	3.82
天　津	13	2.1	3	0.3	1	1	0.03
合　计	3,104	524.6	1,516	318.7	489	1,125	368.1

注：①数据截至2011年6月30日。

②1 $MtCO_2e$=1,000,000tCO_2e。

资料来源：同表3–5。

表 3-16 我国清洁发展机制项目行业分布状况

项目类型	国家批准		注册		核证减排量签发		
	项目数（个）	预期年减排量（$MtCO_2e$）	项目数（个）	预期年减排量（$MtCO_2e$）	项目数（个）	批次（笔）	预期年减排量（$MtCO_2e$）
新能源和可再生能源	2,229	277.3	1,220	149.1	377	738	59.8
节能和提高能效	521	83.7	109	22.8	45	90	12.9
甲烷回收利用	216	52.0	107	28.4	28	64	6.9
燃料替代	42	13.5	29	30.0	16	47	14.5
N_2O分解	32	24.6	27	21.0	12	48	43.5
原料替代	31	4.9	5	1.2	0	0	0.0
HFC-23分解	11	66.8	11	65.7	11	138	230.4
垃圾处理	9	1.1	4	0.2			0.0
资源回收利用	7	0.3	0	0.0	0	0	0.0
造林再造林	4	0.1	3	0.1	0	0	0.0
SF_6回收利用	2	0.3	1	0.2			0.0
合　计	3,104	524.6	1,516	318.7	489	1,125	368.1

注：①数据截至2011年6月30日。

② 1 $MtCO_2e$=1,000,000tCO_2e。

资料来源：同表3-5。

3.2.5 我国开发与实施清洁发展机制项目的成功经验

从全球清洁发展机制项目的发展状况来看，我国是清洁发展机制项目发展最快、数量最多，实施情况优良的国家。我国清洁发展机制项目得以如此快速、顺利发展，主要得益于以下几个方面[1]：

1 中国清洁发展机制基金管理中心. 气候变化融资［M］. 北京：经济科学出版社，2011：274-276.

3.2.5.1　政府高度重视、大力支持清洁发展机制工作

（1）成立专门管理机构。我国自清洁发展机制实施伊始就成立了专门的政府管理机构——由国家发展和改革委员会、外交部、科学技术部、财政部、国家环境保护总局（现为环境保护部）、农业部和中国气象局等七个部委组成的清洁发展机制项目审核理事会，具体负责我国清洁发展机制项目的审核、批准事宜，同时在国家发展和改革委员会设立应对气候变化办公室（现为气候变化司），具体负责指导我国项目的开发实施、受理项目审批，以及作为国家指定联系机构与联合国清洁发展机制执行理事会沟通等事宜。

（2）出台专门管理办法。为推进项目的顺利实施，早在2004年6月30日，即我国清洁发展机制项目实施起步阶段，政府就颁布了《清洁发展机制项目运行管理暂行办法》，以规范我国项目开发流程。经过一段时间的实践，结合项目实际运行中遇到的新情况、新问题，又先后于2005年10月12日正式颁布《清洁发展机制项目运行管理办法》，于2011年8月3日对管理办法进行了修订。相关制度的出台、完善，规范了我国项目开发和国家审批流程，保证了项目实施的公开、透明、公平，维护了项目有序实施。

（3）开展全国范围的专业能力建设。鉴于清洁发展机制的全新性和高度的国际性，在其实施初期，政府就在全国范围内大力开展能力建设，通过对各级政府、企业、相关从业机构等不同层面的人员开展专题培训、讲座、研讨和经验交流等活动，使“清洁发展机制”理念得到广泛推广，一批企业、从业机构迅速掌握了必备的专业技术能力，并在实践中快速提高。

（4）设定企业与国家共同分享清洁发展机制碳交易收入制度。为使清洁发展机制碳交易对国家应对气候变化和节能减排工作发挥更大作用，考虑到温室气体减排资源的公共性，我国对清洁发展机制碳交易收入设定了国家与企业共同分享制度，并成立了中国清洁发展机制基金及其管理中心，统筹收取、管理和使用国家收入部分的资金。中国清洁发展机制基金的资金专门用于支持包括清洁发展机制在内的国家应对气候变化工作。这一制度，使得清洁发展机制这一原本为国外企业与国内企业之间的合作机制，提升到国家层面，便于政府集中一部分资源，统筹推进国家应对气候变化和节能减排整体战略，放大了清洁发展机制对促进我国可持续发展的作用。

3.2.5.2 我国经济发展为实施清洁发展机制项目提供了机遇

近年来，我国政府高度重视环境友好型、资源节约型社会建设，大力推进节能减排工作，促进经济结构转型。一方面淘汰落后产能，提升产业技术水平，降低能耗；另一方面，积极鼓励新能源和可再生能源发展。这些活动是清洁发展机制的优质项目资源，为我国清洁发展机制项目发展提供了难得的机遇。同时，实施清洁发展机制项目也可推动企业技术革新、产业转型。另外，我国工业类型齐全，还拥有较多HFC-23分解、N_2O分解等规模较大的清洁发展机制项目。

3.2.5.3 我国丰富的自然资源促进清洁发展机制项目多元化

我国幅员辽阔，拥有丰富的风电、水电、太阳能、地热能、潮汐能等自然资源。这使得我国清洁发展机制项目资源丰富，类型多元化。

3.2.5.4 咨询公司是我国清洁发展机制项目发展的重要引擎

由于清洁发展机制项目遵循的国际规则和程序复杂，又以英语作为工作语言，参照的法律也与国际贸易法相关，我国绝大多数项目业主难以依靠自身的力量有效地开发、实施清洁发展机制项目。我国在推广清洁发展机制初期就充分意识到了这一问题，非常重视专业咨询机构的作用，在国内迅速培养了一批专业化的咨询机构。这些咨询机构在市场动力驱动下，在全国各地寻找、开发项目，积极向潜在的项目业主推介清洁发展机制的概念，提供一站式服务，极大地推动了清洁发展机制在我国的普及和发展。据初步统计，目前在我国从事清洁发展机制项目服务的咨询机构已达百余家之多，而且它们多为中小型民营企业，这对于促进国家经济发展结构转型，支持中小企业发展都有着重要意义。

此外，我国政府还积极培育国内的独立第三方核查核证机构。目前，我国获得联合国清洁发展机制执行理事会认可的核查核证机构已达4家。这一本土化工作，有效消除了国外核查核证机构因语言、文化差异等因素影响项目核证、核查进度的情况。

第4章

清洁发展机制的主要作用及其问题

4.1 清洁发展机制的主要作用

清洁发展机制被普遍认为是一种双赢机制。理论上，发达国家通过这种项目级的合作，能以远低于其国内减排成本实现其在《京都议定书》下的减排承诺，节约大量资金，并且可以通过这种方式将技术、产品、服务等一并向发展中国家输出；发展中国家通过这种合作，可以获得清洁发展理念、技术，并获得一定数量的减排资金，从而促进本国的经济发展和环境保护，促进本国的可持续发展。通过几年的实践，清洁发展机制项目对全球共同应对气候变化的贡献主要表现在以下几方面[1]:

4.1.1 帮助发达国家降低减排成本

清洁发展机制作为当前全球唯一有效联系发达国家和发展中国家共同应对气候变化的成功机制，使发达国家在发展中国家的帮助下，能够以较低成本实现其温室气体减排义务。当前发达国家在其国内减少1吨二氧化碳当量的成本高达70～100美元[2]，而发达国家在发展中国家清洁发展机制一级市场购买1吨二氧化碳当量的减排量只需5～15美元。通过清洁发展机制项目实施，发展中国家仅在2010年就帮助发达国家节约减排资金达110亿～650亿美元，累计已达420亿～2,440亿美元。从某种意义上说，我国通过清洁发展机

1 谢飞，孟祥明，胡烨. 清洁发展机制：撬动发展中国家低碳经济杠杆[N]. 中国财经报，2010年1月21日（第4版）.

2 中国清洁发展机制的冲动与尴尬. http://news.sina.com.cn/c/sd/2009-12-23/140519322122.shtml.

制项目已为发达国家节约了相当可观的履约成本。

4.1.2 为发展中国家低碳发展提供资金支持

据测算，在2010年，清洁发展机制项目的核证减排量交易为发展中国家直接带来了超过12亿美元的资金；从清洁发展机制项目实施至今，累计带来资金已超过50亿美元。其中，为我国累计带来的资金约20亿美元。同时，通过清洁发展机制项目的开发、建设和运行等，间接撬动的融资资金达数百亿美元。这在一定程度上促进了发展中国家的低碳发展，特别是在全球金融危机形势下，这些由清洁发展机制直接和间接带来的资金对发展中国家社会和经济发展的贡献显得更为突出，使不少项目企业减轻了资金压力，企业得以继续运营。

4.1.3 为发展中国家可持续发展提供新理念

清洁发展机制有效地把环境保护活动市场化，推动各方积极开展相关活动，也为发展中国家探索如何借助市场手段解决其他环境问题提供了新思路和实践参考。同时，发展中国家企业通过清洁发展机制项目实施、与联合国清洁发展机制执行理事会和指定经营实体沟通等，学习和吸收了国际先进企业理念和管理经验，提高了企业的科学化、规范化和精细化管理水平，帮助企业健康发展。我国通过历时五年的实践，使很多企业对节能减排和低碳发展有了更深入的认识，对发达国家利用市场手段（如碳市场机制）推动节能减排工作有了较全面的了解。目前，国家已确定“五省八市”作为低碳发展试点[1]，多途径探索国内低碳发展路线，并已明确提出在“十二五”规划期间健全节能市场化机制，逐步建立碳排放交易市场[2]。清洁发展机制对帮助我国开展这些活动无疑是起到了很好的铺垫作用。

4.1.4 为发展中国家培养国际化环保队伍

由于在清洁发展机制项目开发、实施过程中，发展中国家的项目业主、

1 国家发展和改革委员会文件. 国家发展改革委关于开展低碳省区和低碳城市试点工作的通知. http://www.sdpc.gov.cn/zcfb/zcfbtz/2010tz/t20100810_365264.htm.

2 中华人民共和国国民经济和社会发展第十二个五年规划纲要. http://news.xinhuanet.com/politics/2011-03/16/c_121193916_12.htm.

咨询机构以及指定经营实体等均需按照国际统一规则行事，这客观上促进了发展中国家相关工作人员素质和能力的提高，对于推进本国其他环境保护工作意义重大。这点在我国清洁发展机制项目企业中有很好的体现。不少企业反映，通过开发和实施清洁发展机制项目，企业项目参与人员开阔了眼界，能力和知识水平都得到大幅提升。

另外，在实施清洁发展机制项目中，国内有一批认证机构的能力得到了培养和提高，目前已有中环联合认证中心有限公司、中国质量认证中心、中国船级社质量认证公司和赛宝认证中心四家国内认证机构获得了指定经营实体的资格。这些机构获得联合国清洁发展机制项目执行理事会的认证，除了有利于促进我国清洁发展机制项目实施以外，还将有利于提升我国量化自行开展的节能减排工作的水平，并使之与国际接轨。

4.2　清洁发展机制的主要问题

4.2.1　全球发展严重不均衡

从全球清洁发展机制项目在各国的发展情况（包括注册项目数、预期年减排量和核证减排量签发）可以看出，中国、印度和巴西三国的市场份额占71%以上，形成绝对垄断地位。这使得清洁发展机制这一个联合国主导下的全球机制不能惠及广大发展中国家，促进更多发展中国家的可持续发展，同时也使得中国、印度、巴西等国成为众矢之的。这既对清洁发展机制的健康发展极为不利，也影响其他发展中国家的积极性，进而影响全球共同应对气候变化目标的实现。造成这一现象的主要原因包括：（1）清洁发展机制项目实施周期长，对企业的长期稳定发展要求较高，这使得国际买家更愿意在一些能力强、政局稳定的国家做项目。（2）相对于大的发展中国家而言，小的发展中国家和小岛屿国家的大型企业较少，温室气体排放基数较低，不易开发出大型的清洁发展机制项目。从降低交易成本考虑，国际买家更愿意到大的发展中国家购买大型清洁发展机制项目。（3）清洁发展机制项目规则烦琐、复杂，使得一些能力较弱的国家不易于开展清洁发展机制[1]。

1　孟祥明，冯超，谢飞. 全球CDM市场发展及其面临的挑战［J］. 经济研究参考，2009，2217（17）：5–9.

4.2.2 与全球应对气候变化需求相距甚远

从全球清洁发展机制发展状况可以看出，因各种因素制约，清洁发展机制无论是实际效果还是预期效果，均与全球应对气候变化的需求相距甚远。例如，截至2011年6月30日，全球获得核证减排量签发的项目为1,088个，仅占全球已批准注册项目（3,368个）的32.3%。我国国家批准的3,104个项目中，只有1,516个项目注册成功，占国家批准项目的48.8%；只有489个项目获得核证减排量签发，占注册成功项目的32.2%，占国家批准项目的15.8%。制约发展的因素主要包括：（1）清洁发展机制程序烦琐，方法学适用面窄，且更新频繁。（2）联合国清洁发展机制项目执行理事会工作效率低下，导致大量项目的注册和核证减排量签发申请积压。（3）指定经营实体的能力严重不足。截至2011年6月30日，全球仅有38家咨询机构获得指定经营实体资质，除2003年全球清洁发展机制项目刚启动时联合国清洁发展机制项目执行理事会连续批准的11个机构外，后期发展极为缓慢。（4）清洁发展机制的决策程序仍缺乏透明度，相关规定的众多环节为联合国清洁发展机制项目执行理事会留有裁量权，比如2010年出台的核证核查指导手册中，对众多参数的选取、操作的细节等描述仍较为含糊[1]。

4.2.3 只能作为发展中国家应对气候变化活动的有益补充

从当前全球清洁发展机制市场发展状况来看，清洁发展机制一级市场的规模有限，交易价格较低，因此其为发展中国家带来的资金量有限。据世界银行发布的《2011年全球碳市场发展现状与趋势》[2]分析，自2005年全球碳市场快速发展以来，随着全球碳市场发展成熟与稳定，与发展中国家直接相关的清洁发展机制一级市场交易额占全球碳市场交易总额的比重逐步降低，由2005年的23.6%下降至2010年的不足1.1%，累计为发展中国家带来的资金总额仅为265亿美元，只占全球碳市场累计交易总额的5.0%。且受2012年后《京都议定书》第二承诺期政策不确定性以及欧盟等发达国家对清洁发展机制的打压，自2007年清洁发展机制一级市场交易额达到74亿美元以后，连续

1 谢飞，孟祥明，刘淼. 全球碳市场：冷热不均 寻求突破[N]. 中国财经报，2010年6月24日（第4版）.

2 Carbon Finance at World Bank. State and Trends of the Carbon Market 2011: 9.

三年呈现两位数下降，2010年全球清洁发展机制一级市场更是骤降44.4%，仅为15亿美元，已经低于2005年清洁发展机制起步时的交易额（26亿美元）（2005年以来全球碳市场发展情况详见表4-1）。故在现有机制不改革的情形下，清洁发展机制难以像欧盟和美国等发达国家所建议的那样，成为发展中国家应对气候变化的一个主要资金来源[1]。

表 4-1　2005~2010年全球碳市场交易额统计情况　单位：亿美元

年份	清洁发展机制一级市场	全球碳市场	比例（%）
2005	26	110	23.6
2006	58	312	18.6
2007	74	630	11.7
2008	65	1,351	4.8
2009	27	1,437	1.9
2010	15	1,419	1.1
合　计	265	5,259	5.0

资料来源：Carbon Finance at World Bank. State and Trends of the Carbon Market 2011.

在发展中国家应对气候变化资金需求方面，根据联合国气候变化框架公约秘书处2007年发布的技术报告测算，到2030年，发展中国家每年用于应对气候变化的资金需求达1,200亿 ~ 1,630亿美元[2]。哥本哈根协议承诺，2010 ~ 2012年，发达国家每年为发展中国家提供300亿美元，2013 ~ 2020年，每年为发展中国家筹集1,000亿美元[3]。联合国秘书长气候变化融资高级顾问团对每年1,000亿美元的资金来源给出的建议是，通过碳市场每年筹集约100亿美元。

比较发展中国家的气候融资需求和发展中国家通过碳市场能够获取资金的情况可以看到，如果没有大的变革，发展中国家从碳市场可以获取的资金极为有限，远不能满足其应对气候变化活动的需求。

1　谢飞，许明珠，孟祥明. 欧盟推出最新气候变化战略［N］. 中国财经报，2010年4月8日（第4版）.

2　中国清洁发展机制基金管理中心. 气候变化融资［M］. 北京：经济科学出版社，2011：56.

3　UNFCCC, Decisions adopted by the Conference of the Pargties. http://unfccc.int/resource/docs/2009/cop15/eng/11a01.pdf, P6.

全球清洁发展机制项目关键指标情况统计

一个清洁发展机制项目从注册成功到核证减排量获得签发的用时情况，以及从一个减排量核证期结束到该批核证期的减排量顺利获得签发的用时情况（核证减排量签发用时），是衡量项目速度的两个关键性指标。

（1）从项目注册成功到核证减排量获得签发的用时情况。通过对截至2011年6月30日全球已获得核证减排量签发的1,088个项目进行分析可见，当前，这些项目从注册成功到第一批核证减排量获得签发的平均用时为570天。随着全球注册项目申请的逐步增多，这一用时逐步增加。在2010年下半年及2011年上半年联合国清洁发展机制执行理事会加快项目审批速度，2011年4月底前，积压时间较久的项目获得签发，这期间签发的这些项目从注册到第一批核证减排量顺利获得签发的用时高于平均值的项目仍较多，但自2011年5月后，随着积压项目受理基本结束，从注册到第一批核证减排量签发的用时在明显缩短。具体情况详见图4-1。

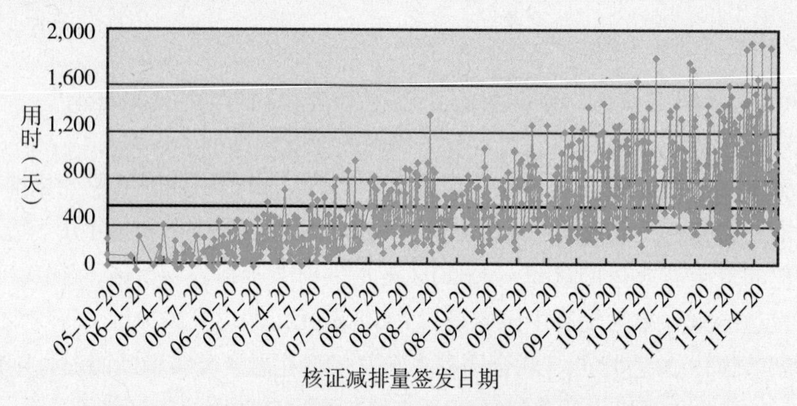

图 4-1　项目从注册成功到核证减排量获得签发用时情况

资料来源：根据联合国气候变化框架公约清洁发展机制执行理事会网站数据整理而得。

（2）从减排量核证期结束到该批核证减排量获得签发的用时情况。从项目某一减排量核证期结束到该批核证减排量获得签发所需时间通常叫核证减排量签发用时。通过对全球已签发的2,789批核证减排量进行统计可见，核征减排量签发用时平均为302天。随着签发申请增多，核证减排量签发用时在逐步增加。但在2010年下半年和2011年上半年，随着积压的签发申请被集中处理，自2011年6月起，核证减排量签发用时降至363天。从核证减排量签发用时的统计来看，在101～400天的签发批次占已签发总批次的69.8%。全球清洁发展机制项目核证减排量签发用时情况详见图4-2和图4-3。

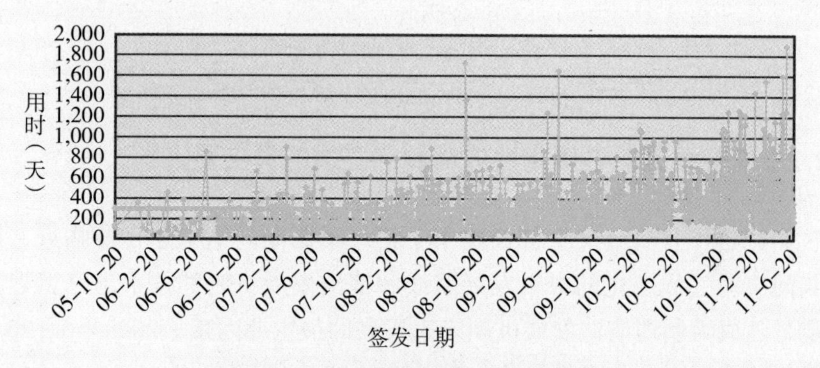

图 4-2　核证减排量签发用时情况

资料来源：同图4-1。

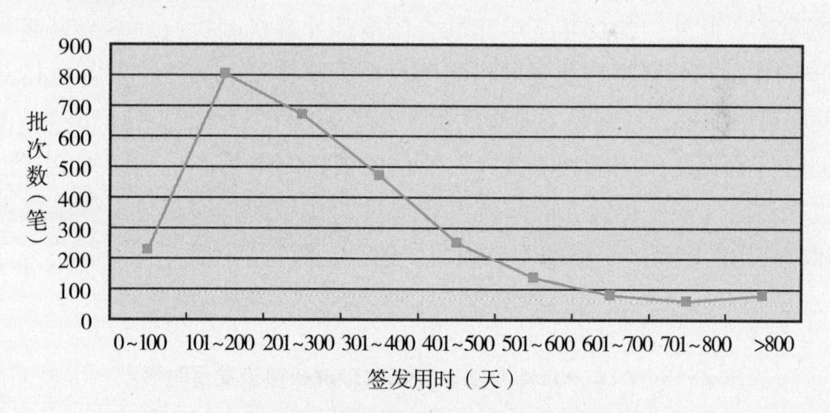

图 4-3　签发用时各时间段内的签发批次情况统计

资料来源：同图4-1。

通过对关键指标分析可见，尽管在2010年联合国清洁发展机制项目执行理事会采取了各种措施，努力加快审批速度，使得新注册项目数量和核证减排量签发批次有显著增加，但由于项目实施过程中涉及指定经营实体对项目情况进行核证、出具核证报告、向联合国清洁发展机制执行理事会递交核证减排量签发申请、联合国清洁发展机制执行理事会审核签发申请等多个环节，因此项目整体用时仍在增加。到2010年底，一个项目从注册成功到第一批核证减排量获得签发的平均用时已长达2年，项目的核证减排量签发平均用时长达1年。如此长的用时对项目的规模化发展极为不利。

4.2.4 未来政策的不确定性成为制约发展的最主要因素

2012年12月31日《京都议定书》第一承诺期的结束期日益临近，而全球共同应对气候变化的未来发展路线和清洁发展机制的未来出路迟迟不能确定，已成为制约清洁发展机制未来发展的最主要因素。2009年6月25日欧盟修订的欧盟国会及欧盟委员会指令2003/87/EC明确指出，作为全球最大的碳市场和清洁发展机制项目减排量的最大购买方，欧盟排放权交易体系自第三期起只接受来自于最不发达国家的，新注册的清洁发展机制项目产生的减排量[1]。此外，2010年下半年联合国清洁发展机制执行理事会突然暂停对所有HFC-23分解类清洁发展机制项目核证减排量的签发，西方机构对清洁发展机制项目，特别是工业废气类（主要为HFC-23分解、己二酸类N_2O分解）项目的质疑和抨击，使各方对清洁发展机制的信心受到极大的打击。欧盟在坎昆气候变化大会前高调宣布拟自2013年起禁用来自工业废气类的国际碳信用的决议，更是给本已脆弱的清洁发展机制体系雪上加霜。

1 Directive 2003/87/EC of the European Parliament and of the Council of 23 April 2009, http://eur-lex.europa.eu/LexUriServ/LexUriServ.do?uri=OJ:L:2009:140:0063:0087:EN:PDF, P24.

第5章

清洁发展机制的展望

虽然，未来政策不确定性以及部分发达国家对发展中大国的清洁发展机制项目实施的抵制与阻挠，导致当前清洁发展机制陷入低谷，包括中国机构在内的众多机构也对2012年后的项目实施徘徊和犹豫。但欧盟在2009年6月25日修订的欧盟国会及欧盟委员会指令2003/87/EC中明确指出，在全球新的气候变化协议未达成前，欧盟排放交易体系可以在2015年3月31日前，继续接受2008～2012年欧盟排放交易体系下的合规项目在2012年前产生的减排量；还可以继续接受在2012年12月31日前注册成功的合规项目在2013年后产生的减排量[1]。这表明在2012年后，欧盟将在一定条件下继续使用清洁发展机制项目产生的减排量。

2010年底的坎昆气候变化大会上，联合国气候变化公约各缔约方通过的一揽子协议——《坎昆协议》中，各方承诺在《京都议定书》第一和第二承诺期间不断档，将基于现有的市场机制培育更多市场手段[2]，并通过《关于清洁发展机制的进一步指南》，对清洁发展机制的进一步改革提出明确要求。这都为清洁发展机制的未来发展带来曙光[3]。

2011年德班气候变化大会上通过的一揽子协议中也明确《京都议定书》第二承诺期自2013年1月1日开始，至2017年12月31日或2020年12月31日结束。此举坚持了我国等发展中国家一贯坚持的“双轨制”谈判路线，保住了《京都议定书》第二承诺期，也从某种程度上保证了我国参与最广，对我国

1 UNFCCC, Decisions adopted by the Conference of the Pargties. http://unfccc.int/resource/docs/2009/cop15/eng/11a01.pdf, P6.

2 UNFCCC, Report of the Conference of the Parties serving as the meeting of the Parties to the Kyoto Protocol on its sixth session, held in Cancun from 29 November to 10 December 2010, P4.

3 孟祥明，李春毅，谢飞. 2012年后，碳市场和CDM不会消失［N］. 中国能源报，2011年1月24日（第6版）.

可持续发展贡献最大和目前运行最成功的清洁发展机制[1]。

虽然目前对《京都议定书》第二承诺期清洁发展机制的众多技术问题尚未确定，欧盟也在加大对中国等新兴经济体清洁发展机制的限制，但可以明确清洁发展机制在《京都议定书》第二承诺期将继续存在。这为我国众多从事该项工作的项目业主、咨询机构等坚定了信念。即使哪天因各方争执不下而导致清洁发展机制夭折，各种形式的碳市场和促进发展中国家参与应对气候变化的各种市场机制也将存在，因为市场机制在应对气候变化中的有效性已被各方广泛认可，并被事实多次证明。

1 Europe has a planto keep CDM if Kyoto Protocol expires. http://www.rechargenews.com/hardcopy/article274560.ece.

下　篇

中国清洁发展机制基金

第6章

中国清洁发展机制基金的由来

6.1 清洁发展机制项目国家收入

清洁发展机制，是一个项目级的合作，即发达国家的温室气体减排量购买机构与发展中国家清洁发展机制项目企业之间的合作，前者向后者提供资金，后者向前者提供减排量。发展中国家政府在清洁发展机制项目的开发、实施过程中，得不到任何资金收入，而发展中国家政府恰恰受国家发展阶段所限缺乏开展应对气候变化工作的资金。这可谓清洁发展机制设计上的一个不足。毫无疑问，如果发展中国家政府能够从清洁发展机制合作中获得一定的资金收入，用以支持本国国家层面的应对气候变化工作，该机制对促进全球共同应对气候变化无疑将是有效的。

我国政府在开展清洁发展机制项目初期就认识到，温室气体减排容量像矿产资源、环境资源等一样，是一种公共资源，应归政府所有，由此产生的温室气体减排量交易收入也应归政府和项目企业共同分享。归政府所有部分称作国家收入。而我国拥有较多减排规模巨大的HFC-23及N_2O分解两类工业废气分解项目这一事实，则进一步推动了国家收入这一概念在我国的提出。

HFC-23和N_2O均是相关化工产品生产过程中产生的废气，在开展清洁发展机制项目之前，均不经处理直接排空。但HFC-23的全球升温潜能（GWP）是二氧化碳的11,700倍，N_2O的全球升温潜能是二氧化碳的310倍，对全球气候变暖影响巨大。这两类工业废气经分解开发成清洁发展机制项目后的共同特点是，可以产生巨大的减排量，但相对于风电、水电、能效提高等项目而言，其减排成本较低。这两类项目主要集中在我国、印度等少数几个国家。国际社会曾一度担心如果这两类项目没有任何约束，可能导致企业因开展HFC-23或N_2O分解清洁发展机制项目，获得丰厚的减排量收入，而刺激其不顾主产品的市场情况进行恶意生产。为了打消国际社会的这种担忧，表明我国的负责任态度，我国政府决定对这两类清洁发展机制项目收取较高比例的国家收入。

2004年6月30日，我国出台的《清洁发展机制项目运行管理暂行办法》规定："项目因转让温室气体减排量所获得的收益归中国政府和实施项目的企业所有。分配比例由中国政府确定，确定前归该企业所有。"2005年10月，在颁布《清洁发展机制项目运行管理办法》时进一步完善为："鉴于温室气体减排资源归中国政府所有，而由具体清洁发展机制项目产生的温室气体减排量归开发企业所有，因此，清洁发展机制项目因转让温室气体减排量所获得的收益归中国政府和实施项目的企业所有。分配比例如下：

（1）氢氟碳化物和全氟碳化物类项目，国家收取转让温室气体减排量转让额的65%；

（2）氧化亚氮类项目，国家收取转让温室气体减排量转让额的30%；

（3）本《办法》第四条规定的重点领域以及植树造林项目等类清洁发展机制项目，国家收取转让温室气体减排量转让额的2%。

中国政府从清洁发展机制项目收取的费用，用于支持与气候变化相关的活动。"

2011年8月3日，根据我国清洁发展机制项目的实施情况，我国政府对《清洁发展机制项目运行管理办法》进行了修订，颁布了《清洁发展机制项目运行管理办法（修订）》，对不同类型项目的国家收入比例进行了调整，具体规定如下：

清洁发展机制项目因转让温室气体减排量所获得的收益归国家和项目实施机构所有，其他机构和个人不得参与减排量转让交易额的分成。国家与项目实施机构减排量转让交易额分配比例如下：

（1）氢氟碳化物（HFC）类项目，国家收取温室气体减排量转让交易额的65%；

（2）己二酸生产中的氧化亚氮（N_2O）项目，国家收取温室气体减排量转让交易额的30%；

（3）硝酸等生产过程中的氧化亚氮（N_2O）项目，国家收取温室气体减排量转让交易额的10%；

（4）全氟碳化物（PFC）类项目，国家收取温室气体减排量转让交易额的5%；

（5）其他类型项目，国家收取温室气体减排量转让交易额的2%。

国家从清洁发展机制项目减排量转让交易额收取的资金，用于支持与气候变化相关的活动，由中国清洁发展机制基金管理中心依据《中国清洁发展机制基金管理办法》收取。

6.2 中国清洁发展机制基金的提出

为贯彻落实《清洁发展机制项目运行管理办法》有关规定，充分发挥清洁发展机制项目产生的国家收入的作用，有效支持我国应对气候变化工作，2006年，财政部牵头，联合国家发展和改革委员会、外交部、科学技术部、环境保护部、农业部和中国气象局，向国务院申请设立中国清洁发展机制基金及其管理中心。该申请很快获得国务院批准。

基金在申请过程中，充分借鉴了社保基金、中比基金、全球环境基金（Global Environment Facility）、英国碳基金（Carbon Trust）等成熟基金的资金管理模式，顺应国家应对气候变化工作快速发展和气候变化国际合作新形势的需要，并充分考虑核证减排量的国际购买方、世界银行、亚洲开发银行等国际组织的有益建议，对基金的业务和运作提出了如下设想：

第一，基金资金来源多样化。基金的主要业务是代表我国政府，收取、管理和使用我国政府从清洁发展机制项目中获得的国家收入，并实现此收入的保值和增值。同时，基金还将积极吸收和利用国内外资源，探索开发新的合作模式，为我国应对气候变化工作争取更广泛的资金支持。

第二，基金为我国应对气候变化工作提供多方位服务。基金将以其收取的国家收入和筹措的各种融资，为国家应对气候变化整体工作服务，包括：（1）支持国家应对气候变化相关的能力建设；（2）支持国家提高应对气候变化的公众意识；（3）促进能效提高和节能；（4）促进可再生能源的开发和利用；（5）促进其他具有显著的减缓气候变化效益的活动；（6）促进气候变化的适应；（7）支持基金可持续运行的金融活动等。

第三，基于既体现国家政策导向，又可以在广阔市场上与商业合作伙伴开展直接合作的考虑，基金定位为按照社会性基金模式管理的政策性基金。基金将在政府的指导和支持下，开展广泛的国际国内合作，引导、推动市场减排，探索发展基于市场的长效机制，支持行业低碳发展和新兴低碳产业发展。

6.3 设立中国清洁发展机制基金的意义

中国清洁发展机制基金作为一个创新的应对气候变化资金机制和行动机制，对于提升清洁发展机制对我国应对气候变化工作的作用，以及我国应对气候变化国际形象具有重要意义。

第一，基金把清洁发展机制这一国际合作从企业层面提升到国家层面，加强了清洁发展机制项目在我国应对气候变化工作中的作用。该基金是一个国家级的应对气候变化专项基金，直接在国家层面服务于我国的应对气候变化工作。特别是，在现阶段“发展”仍是我国的首要任务的情况下，通过基金集中一部分资金，专门用于支持国家层面的应对气候变化的政策研究、规划制定、产业活动等，可以促进我国系统、有序地开展气候变化履约行动和落实国家应对气候变化方案。

第二，作为一个创新的融资机制，基金通过发挥种子资金和引导资金作用，可以撬动和引导大批的社会资金进入我国应对气候变化领域，支持国家应对气候变化工作，支持我国应对气候变化和节能减排工作产业化、市场化和社会化，推动我国的低碳发展。

第三，基金通过基金管理中心与基金资助项目的实施机构、国际机构的广泛交流与合作，促进了我国应对气候变化工作的国际合作。

第 7 章

中国清洁发展机制基金的建立及其治理结构

7.1 中国清洁发展机制基金的建立

7.1.1 中国清洁发展机制基金的筹建

基金的筹建和设立受到我国政府高层的高度重视。财政部谢旭人部长、国家发展和改革委员会解振华副主任、财政部廖晓军常务副部长、朱光耀副部长、原财政部纪检组长贺邦靖等领导多次过问并指导相关工作。

2005年10月，财政部牵头，联合国家发展和改革委员会、外交部和科学技术部，启动了中国清洁发展机制基金筹建工作。

中国政府建立基金的举措，在国际社会引起强烈反响，一些国际发展组织、国际金融组织和外国政府机构，对于与基金开展合作表示了浓厚兴趣和强烈愿望。世界银行和亚洲开发银行为基金的筹建和初期运行的能力建设提供了大力支持。

2005年12月，财政部代表我国政府同世界银行签署了我国第一批、两个HFC-23分解类清洁发展机制项目温室气体减排量购买协议，协议金额约5亿美元。经协商，世界银行同意从这两个项目第一批核证减排量交易所对应的国家收入部分中，提前支付630万美元，用作基金筹建及初期运行的启动资金。这是基金拥有的第一笔资金。

2006年5月，亚洲开发银行为基金提供了一个60万美元的技术援助项目，用于帮助基金的筹建工作，包括设计机构治理架构、编写基本规章制度和加强机构能力等。2008年10月，亚洲开发银行又为基金提供了一个80万美元的技术援助项目，用于帮助基金在建立初期的能力建设，包括建立一系列的规章制度和管理手册，设计、建立一系列信息管理系统，开展人员培训和有关基金的国内外宣传活动等，为基金的机构能力加强、初期运营、开展业务、提高国内外知名度，做出了重要贡献。

7.1.2 中国清洁发展机制基金的设立

在多部委通力合作下，经过历时一年的准备，2006年5月，财政部、国家发展和改革委员会、外交部和科学技术部向国务院呈交了关于建立中国清洁发展机制基金及其管理中心的联合请示。国务院对成立该基金高度重视，迅速做出反应，在征求其他部门的意见和建议后，于2006年8月由国务院主要领导亲自批复，批准建立中国清洁发展机制基金及其管理中心。

基金管理中心的建立得到中央机构编制委员会办公室的大力支持。2007年3月，基金管理中心在国家事业单位登记管理局完成注册登记，成为近年来为数极少的新批准的编制机构之一。2007年4月在国家外汇管理局等部门的大力支持下，中国清洁发展机制基金专用账户正式启用。2007年4月，基金审核理事会召开第一次会议，确定国家发展和改革委员会为基金审核理事会主席，财政部为副主席，并正式启动基金相关制度的编制工作。同月，基金业务开始运营。2007年4月13日，第一笔清洁发展机制项目国家收入进入基金账户。

7.1.3 中国清洁发展机制基金的启动

2007年11月9日，中国清洁发展机制基金及其管理中心启动仪式在京举行，财政部谢旭人部长、李勇副部长、朱光耀副部长，国家发展和改革委员会解振华副主任，外交部张业遂副部长，亚洲开发银行黑田东彦行长等出席启动仪式并发表重要讲话。

财政部谢旭人部长在启动仪式上指出，中国清洁发展机制基金的设立是财政部深入贯彻落实科学发展观，创新投融资机制，加大资金支持力度的具体体现。财政部将充分发挥中国清洁发展机制基金的“孵化器”作用，大力扶植和发展节能减排项目，搭建应对气候变化和节能减排领域的国际合作与行动平台，推动国家可持续发展战略和全球应对气候变化行动的开展。同时还将通过中国清洁发展机制基金与财政资金的有效结合，引导并撬动社会力量的广泛参与，实现公共部门与私营部门的资源整合，为

财政部谢旭人部长与国家发展和改革委员会解振华副主任共同启动中国清洁发展机制基金

构建资源节约型、环境友好型社会贡献力量。

国家发展和改革委员会解振华副主任在讲话中强调，中国清洁发展机制基金的建立在中国应对气候变化和节能减排领域具有里程碑式的意义，随着国际社会对气候变化问题的日益关注，以及中国政府对应对气候变化和节能减排工作的重视程度不断提高，中国清洁发展机制基金将在推动应对气候变化的行动与合作、促进节能减排事业的开展等方面发挥越来越重要的作用。

亚洲开发银行行长黑田东彦出席基金启动仪式并致辞

亚洲开发银行行长黑

田东彦在讲话时表示，清洁发展机制基金是把国际碳市场的收入用于消除国内障碍，实现低碳经济的创新典型。亚洲开发银行将继续与中国清洁发展机制基金开展全面合作，促进中国应对气候变化工作。

中国清洁发展机制基金的启动在国内外受到广泛关注。

7.1.4 《中国清洁发展机制基金管理办法》正式颁布

2010年9月14日，经财政部、国家发展和改革委员会、外交部、科学技术部、环境保护部、农业部、中国气象局部（委、局）务会议通过，报请国务院审批，《中国清洁发展机制基金管理办法》以七部委令的形式正式颁布施行。该办法从管理机构及职责、基金筹集、基金使用、赠款项目管理、有偿使用项目管理等方面对基金业务发展进行了全面界定，为基金全面展开工作提供了政策依据。

7.2 中国清洁发展机制基金的治理结构

7.2.1 宗旨、性质及战略定位

根据《中国清洁发展机制基金管理办法》，基金是由国家批准设立的按照社会性基金模式管理的政策性基金。基金的宗旨是支持国家应对气候变化工作，促进国家可持续发展。基金将在国家可持续发展战略的指导下，支持和落实《中国应对气候变化国家方案》、《节能减排综合性工作方案》等的具体任务。

作为应对气候变化工作的新资源和新机制，中国清洁发展机制基金将政府投入、国际援助、社会资金紧密结合起来，是一个资金合作平台、行动合作平台和信息收集与交流平台。基金通过赠款、股权投资、委托贷款、融资性担保等资助方式，支持能力建设、公众意识宣传、能效提高和新能源发展等活动，配合国家主渠道资金，支持国家应对气候变化工作。同时，基金还将依托政府的支持，通过社会性基金运行模式，与企业、商业金融机构等开展密切合作，规模化地推广成功的示范项目，发挥政策性基金的作用。

7.2.2　治理结构

基金审核理事会和基金管理中心是基金的组织机构。基金审核理事会是关于基金事务的部际议事机构。

基金管理中心是基金的日常管理机构，负责基金的筹集、管理和使用工作。基金管理中心在基金审核理事会指导下开展业务工作。

为保证中国清洁发展机制基金管理中心业务稳定持续发展，完善基金管理中心治理结构，提高决策的科学性，控制风险，增强竞争力，健全重大事项决策程序，基金管理中心还成立了战略发展委员会、风险控制委员会、投资评审委员会。审核理事会和战略发展委员会的外部制衡，风险管理委员会、投资评审委员会的内部制约，项目立项、调查、审批、风险控制和监督各环节环环紧扣、相互制衡，从体制上保证了基金的安全、高效运行。

中国清洁发展机制基金治理架构如图7–1所示。

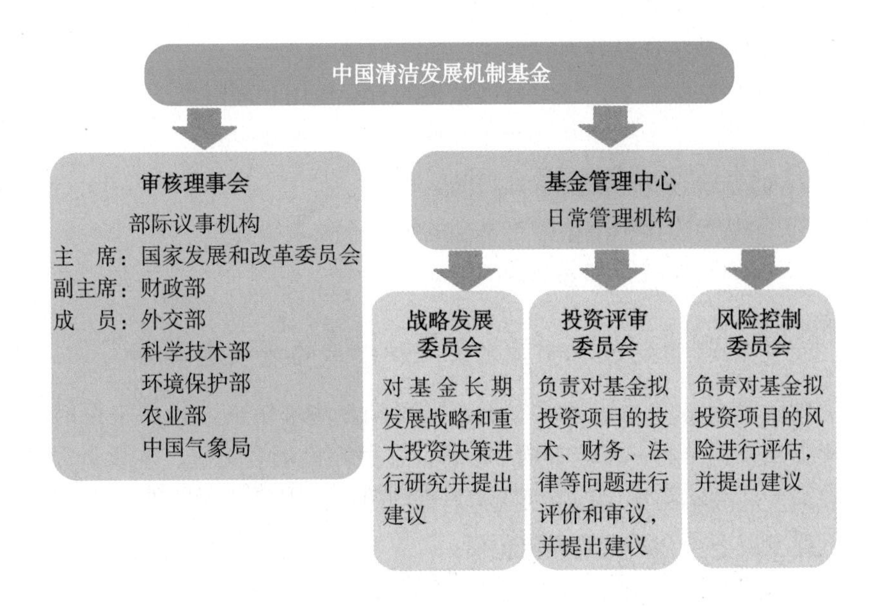

图 7–1　中国清洁发展机制基金治理架构

7.2.2.1　中国清洁发展机制基金审核理事会

基金审核理事会是基金事务的部际议事机构。基金审核理事会由国家发展和改革委员会、财政部、外交部、科学技术部、环境保护部、农业部和国家气象局的代表组成。基金审核理事会设主席和副主席，分别由国家发展和改革委员会与财政部派出代表履行职责。

基金审核理事会负责审核下列事项：

（1）基金基本管理制度；

（2）基金发展战略规划，包括资金使用年度计划；

（3）基金赠款项目和重大有偿使用项目申请；

（4）基金年度财务收支预算与决算；

（5）基金其他重大业务事项。

7.2.2.2 中国清洁发展机制基金管理中心

基金管理中心是基金的日常管理机构，具体负责基金的筹集、管理和使用工作，由财政部归口管理。

基金管理中心主要履行下列职责：

（1）起草基金基本管理制度，制定基金具体运行管理规定；

（2）筹集基金资金；

（3）管理基金资金，组织开展基金的有偿使用和理财活动；

（4）编制并组织实施基金的年度财务收支预算与决算；

（5）监督管理基金所支持项目的运行；

（6）向基金审核理事会报告基金的重大业务事项；

（7）开展其他符合基金宗旨的活动。

7.2.2.3 中国清洁发展机制基金管理中心战略发展委员会

为保证基金业务稳定持续发展，完善基金治理结构，提高决策的科学性，增强基金竞争力，基金管理中心设立了战略发展委员会，主要负责对基金中长期发展战略和重大投资决策进行研究并提出建议。

战略发展委员会的主要职责包括：

（1）对基金管理中心中长期发展战略规划进行研究并提出建议；

（2）对基金管理中心的重大投资、融资方案进行研究并提出建议；

（3）对基金管理中心的重大资本运作、资产经营项目进行研究并提出建议；

（4）对其他影响基金管理中心发展的重大事项进行研究并提出建议；

（5）对以上事项的实施进行审议；

（6）基金管理中心授权的其他事宜。

2009年4月16日，第一届战略发展委员会成立并召开第一次战略发展委员会全体会议，会上选举出由中纪委委员、财政部原纪检组长贺邦靖同志担任战略发展委员会主席，朱光耀副部长担任副主席。

7.2.2.4　中国清洁发展机制基金管理中心投资评审委员会

投资评审委员会是基金管理中心设立的议事工作机构，主要负责对基金管理中心的投资规划、战略提出建议，对基金管理中心拟投资项目进行评价和审议并提出建议。委员会对基金管理中心负责，委员会的投资评审决议作为基金管理中心投资决策的参考依据。

投资评审委员会的主要职责包括：

（1）审议管理中心有偿资金管理战略、评估标准、制度和政策；

（2）审议管理中心有偿资金的投资方向、投资策略及投资限制；

（3）审议管理中心有偿资金使用年度计划；

（4）根据预算、项目可行性研究报告、项目实施方案和尽职调查报告，独立开展基金有偿使用项目的技术和经济评价工作；

（5）对管理中心已投资项目的监管和投资运作情况进行督导；

（6）审议管理中心对已投资项目所做出的处置方案；

（7）审议管理中心对可能出现风险的已投资项目所做出的计提坏账准备方案；

（8）对其他影响基金发展的重大投资事项进行研究并提出建议；

（9）必要时对投资项目进行现场考察；

（10）执行管理中心及《中国清洁发展机制基金管理办法》、《中国清洁发展机制基金有偿使用管理办法》授权的其他事宜。

7.2.2.5　中国清洁发展机制基金管理中心风险控制委员会

风险控制委员会是基金管理中心设立的专门议事机构，负责对基金的系统性风险、体制性风险以及有偿资金使用风险进行监控和管理。

风险控制委员会的主要职责包括：

（1）审议基金风险管理的战略和系统规划。

（2）评估有偿资金使用项目风险，提出风险处置建议。风险控制委员会会议的项目风险评价报告将作为项目评审的必要参考依据，提交投资评审委员会，风险评价报告同时提交主任办公会审议决定。

（3）指导改进和完善基金风险管理组织架构、控制程序、风险处理等方面存在的系统性和制度性问题。

（4）基金管理中心及《中国清洁发展机制基金管理办法》、《中国清洁发展机制基金有偿使用管理办法》授权的其他事宜。

此外，为保证基金管理中心有偿使用项目的开发、评估、实施、监督科学合理，规避风险，基金管理中心建立了专门的专家库。专家库里收纳了政策、行业、财务金融、风险管理和法律等六类专家。当投资评审委员会和风险控制委员会需要时，基金管理中心从专家库中随机抽取专家，以保证专家选取的公平、公正性。

第8章

中国清洁发展机制基金管理中心的主要业务

根据国务院授权及《中国清洁发展机制基金管理办法》规定，基金管理中心的业务主要包括基金资金筹集、使用、管理三个主要方面。

8.1 中国清洁发展机制基金资金的筹集

8.1.1 基金的来源

多渠道筹集基金资金，不断扩大基金资金规模是基金运转的根本。根据《中国清洁发展机制基金管理办法》，基金资金来源主要为以下四方面：

（1）通过清洁发展机制项目转让温室气体减排量所获得收入中属于国家所有的部分，即国家收入；

（2）基金运营收入；

（3）国内外机构、组织和个人捐赠；

（4）其他来源，包括社会融资、财政划拨资金等。

目前，基金资金的最主要来源是国家收入，故以下重点介绍国家收入及其收取。

8.1.2 国家收入

基金管理中心按照国家相关规定开设专用账户用于收取国家收入，并比照财政资金收取和管理国家收入，将其全额纳入基金。

8.1.2.1 国家收入的确定

我国清洁发展机制项目业主在获得由联合国清洁发展机制执行理事会签发的每一批次核证减排量转让收入后，需按照《中国清洁发展机制基金管理办法》和《清洁发展机制项目运行管理办法（修订）》的要求，及时、足额缴纳国家收入。基金管理中心秉持公平、公开、公正、可操作的原则，负责国家收入收取。

国家收入按以下公式计算：

$$\text{国家收入}=\left(\text{核证减排量}-\text{捐赠联合国气候变化适应基金核证减排量}\right)\times\text{交易单价}\times\text{国家收入比例}$$

捐赠联合国气候变化适应基金核证减排量，是指按照联合国清洁发展机制规则,每次签发的核证减排量中捐赠给联合国气候变化适应基金的2%部分。

交易单价，是指在项目向国家申请批准函时，递交的项目业主与减排量购买方签订的减排量转让合同中约定，并由国家发展和改革委员会出具的批准函认可的减排量转让交易单价。

自2011年8月3日起，国家收入比例按照《清洁发展机制项目运行管理办法（修订）》的规定执行。具体比例如下：

（1）氢氟碳化物（HFC）类项目，国家收取温室气体减排量转让交易额的65%；

（2）己二酸生产过程中的氧化亚氮（N_2O）项目，国家收取温室气体减排量转让交易额的30%；

（3）硝酸等生产过程中的氧化亚氮（N_2O）项目，国家收取温室气体减排量转让交易额的10%；

（4）全氟碳化物（PFC）类项目，国家收取减排量转让交易额的5%；

（5）其他类型项目，国家收取温室气体减排量转让交易额的2%。

8.1.2.2　国家收入的收取

（1）国家收入收取流程。

首先，交易信息备案。项目业主在减排量转让交易合同（Emission Reduction Purchase Agreement，ERPA）生效后15个工作日内，向基金管理中心提交合同副本、营业执照复印件、合同双方联系人及联系信息等，用于备案。备案事项发生变更的，项目业主应当自变更之日起15个工作日内以书面形式告知基金管理中心，并提交相关证明文件。

其次，项目业主提交支付确认书。项目业主在每次核证减排量签发和转让后的15个工作日内，以书面形式，向基金管理中心提交加盖公章的《中国清洁发展机制项目国家收入支付确认书》（由基金管理中心制作，并在基金管理中心官方网站提供下载）。

再其次，项目业主支付国家收入。项目业主在每次取得联合国清洁发展机制执行理事会签发的核证减排量转让收入后的15个工作日内，向国家收入专用账户支付国家收入。减排量转让合同有相关约定的，国家收入也可由减排量购买方直接向国家收入专用账户支付。

最后，基金管理中心出具收款凭证。基金管理中心在足额收到本次国家收入和证明文件后15个工作日内，向付款方出具收款凭证。按照减排量转让合同约定，由减排量购买方直接支付国家收入的，基金管理中心在国家收入足额取得后，为项目业主出具缴纳国家收入的证明文件。

项目业主在支付国家收入的过程中应注意下列问题：

第一，国家收入可以以减排量转让合同约定的币种或人民币支付。以人民币支付国家收入的，汇率以本次国家收入结汇的现汇买入价为准，并提交加盖企业财务专用章的银行结汇凭证复印件。

第二，减排量购买方以设备等有形资产或专利技术等无形资产向项目业主抵扣减排量转让合同所规定的交易价款的，项目业主按合同约定价格计算国家收入，于本次核证减排量签发和转让后的15个工作日内以人民币形式支付国家收入。汇率以本次核证减排量签发日中国人民银行公布的减排量转让合同约定币种兑人民币中间价为准，并提交加盖企业财务专用章的汇率证明文件。

第三，交易单价为非固定价格的，项目业主须提交加盖企业公章的本次减排量转让交易的收款收据复印件等证明文件。

（2）处罚。

依据《清洁发展机制项目运行管理办法》（修订）第三十一条规定，项目实施机构在减排量交易完成后，未按照相关规定向国家按时足额缴纳减排量交易额分成的，国家发展和改革委员会依法对项目实施机构给予行政处罚。

8.1.2.3 国家收入的预期规模

根据我国清洁发展机制项目实施现状，在2012年以前，如果国际、国内政策不发生重大变化，我国的清洁发展机制项目能够顺利实施，预计到2012年底，我国累计可获得的总收入约为85亿～100亿美元，国家收入部分预计达19亿美元（折合人民币约122亿元[1]）。其中，HFC-23分解类项目对国家收入的贡献为80%左右，N_2O分解类项目的贡献为12%左右，其他类型项目的贡献预计为8%左右。

至于2012年以后的国家收入情况，将取决于《京都议定书》第二承诺期的国际谈判情况、联合国清洁发展机制规则是否有大的变化，欧盟、日本等国际买家的政策，以及我国国内应对气候变化的政策等。

8.2 中国清洁发展机制基金资金的使用

按照资金使用性质划分，基金资金的使用主要包括两种方式：一是以赠款资助国家应对气候变化工作急需的政策研究、能力建设和宣传教育等活动；二是通过有偿使用项目方式支持能够产生应对气候变化效益的产业活动。

8.2.1 赠款

基金赠款主要用于支持国家应对气候变化的下列相关活动：

（1）与应对气候变化相关的政策研究和学术活动；

1 美元对人民汇率按1∶6.4测算。

（2）与应对气候变化相关的国际合作活动；

（3）旨在加强应对气候变化能力建设的培训活动；

（4）旨在提高公众应对气候变化意识的宣传、教育活动；

（5）服务于基金宗旨的其他事项。

基金赠款不支持营利性活动，不用于行政事业支出，原则上不与其他渠道资金重复。

基金赠款的管理和使用主要包括：

8.2.1.1　组织申报

基金审核理事会在每年的基金预算中，基于应对气候变化工作需要，提出当年基金赠款额度，并确定年度重点支持领域和方向，组织赠款项目申报。

8.2.1.2　项目申请

赠款项目申请人根据基金审核理事会确定的重点领域和方向，按照基金审核理事会要求，统一准备赠款项目申请书，内容包括：申请人基本情况、项目背景资料、项目目标、项目主要内容与活动、项目主要产出、项目执行进度安排、申请资金额、预算安排和其他相关内容。申请书由国务院有关部门或省级发展改革部门向国家发展和改革委员会转报或报送。

赠款项目申请人应是我国境内从事应对气候变化领域工作，具有一定研究或者培训能力的相关机构。

8.2.1.3　项目评审

国家发展和改革委员会在收到基金赠款项目申请书后，统一组织由财务专家、管理专家、技术专家等组成的专家组对申请人的资质，项目申请书的完整性，拟申报项目的必要性、可行性，拟申报赠款金额的合理性等内容进行评审，形成评审意见，提交基金审核理事会审核。基金审核理事会在专家评审意见的基础上，对赠款项目审核后，提出拟批准赠款项目清单及对各项目赠款金额的审核意见，由国家发展和改革委员会会同财政部根据基金审核理事会一致意见联合批准确定年度内基金赠款项目名单及各项目赠款额度。

8.2.1.4 合同签署

基金赠款项目由国家发展和改革委员会会同财政部联合批准后，由国家发展和改革委员会、项目组织申报单位、基金管理中心、赠款项目申请人共同签订赠款项目合同，明确各方权利、义务和违约责任处罚办法。

8.2.1.5 项目实施与监督管理

赠款项目的项目申请人应严格按照赠款项目合同实施项目，并接受国家发展和改革委员会、基金管理中心和项目组织申报单位的指导。国家发展和改革委员会、基金管理中心会同项目组织申报单位负责对赠款项目的实施情况进行监督检查。在赠款项目实施过程中的违规行为，由国家发展和改革委员会与财政部依据有关规定予以处理、处罚。

8.2.2 有偿使用

根据《中国清洁发展机制基金管理办法》的规定，基金通过有偿使用的方式支持有利于产生应对气候变化效益的产业活动。基金有偿使用资金重点支持既能产生显著温室气体减排效益，又能产生良好的经济效益和示范效应的产业活动。同时，充分发挥基金的引导、催化作用，积极探索推动我国应对气候变化和节能减排工作产业化、市场化和社会化的创新融资机制，撬动各方资源共同参与，促进国家低碳发展的目标实现。

8.2.2.1 有偿使用方式

有偿使用方式主要包括股权投资、委托贷款、融资性担保以及国家批准的其他方式。

按照基金管理办法的规定，以股权投资、委托贷款方式开展项目的，其年度累计金额不得超过上年末资产净值的一定比例；基金以融资性担保方式支持项目的，其担保额不得超过基金年度预算确定的限额；基金以股权投资方式支持项目的，不得对投资对象控股，投资所形成股权的退出，应当按照公开、公平、公正和市场化的原则，确定退出方式及退出价格。

有偿使用资金不得从事股票、股票类投资基金、房地产以及期货等金融衍生产品的投资。

8.2.2.2　项目申请

有偿使用项目申请人需是中国境内从事减缓、适应气候变化相关领域业务的中资或中资控股企业。项目申请人应按有关规定向基金管理中心提出项目申请。申请文件包括：

（1）项目申请书；

（2）项目可行性研究报告；

（3）企业近3年经营状况；

（4）企业营业执照；

（5）其他相关材料。

8.2.2.3　项目运作和管理

基金管理中心在预算所确定的有偿使用比例和担保限额内，根据国家经济政策、产业政策和应对气候变化政策等情况，进行有偿使用项目的筛选和评审工作。根据基金管理办法规定，单个项目申请基金资金超过7,000万元的属于重大项目，由基金管理中心报经基金审核理事会审核并取得一致意见后，由国家发展和改革委员会、财政部批准；单个项目申请基金资金在7,000万元以下的属于一般项目，由基金管理中心按照规定程序审批，并在批准后报国家发展和改革委员会、财政部备案。

基金管理中心负责对有偿使用项目的组织实施、监督检查和考核验收。有偿使用形成的各种资产及权益将按照国家有关财务规章制度进行管理。

8.3　中国清洁发展机制基金资金的管理

资金管理主要包括对基金收入、使用和支出的管理，具体包括对各种收入的管理，赠款项目资金支出、有偿使用项目资金使用和本外币现金理财业务资金流的管理。

8.3.1 基金收入管理

基金各种收入均实行应收尽收，全额存入基金指定的银行账户。

8.3.2 赠款项目资金管理

赠款项目资金是指经基金审核理事会审核，由国家发展和改革委员会与财政部联合批准的，基金以赠款形式资助的项目资金部分。

赠款项目实施机构须按照赠款项目的资金管理规定，编制赠款项目资金预算，并对赠款项目资金使用用途进行详细说明，提交至基金管理中心。赠款项目实行合同管理。基金管理中心依据签署的赠款项目合同进行资金拨付。

赠款项目资金按项目实施进度，一般分三期拨付：

（1）赠款项目合同签署后，基金管理中心将经批准的赠款项目预算总额的40%拨付给赠款项目实施机构，作为合同准备金。

（2）实施机构提交项目中期报告（包括资金使用报告），经国家发展和改革委员会、项目组织申报单位和基金管理中心审核同意后，基金管理中心将经批准的赠款资金预算总额的另外40%拨付给赠款项目实施机构。

（3）赠款项目通过国家发展和改革委员会、项目组织申报单位和基金管理中心验收合格且财务审计合规后，基金管理中心于15个工作日内向赠款项目实施机构拨付资金预算的余款。

赠款项目资金应当纳入赠款项目实施机构的财务统一管理，单独核算。涉及政府采购的，按照政府采购有关规定执行。国家另有规定的，从其规定。

赠款项目实施过程中，基金管理中心负责对赠款项目执行的财务状况进行监督，并将结果及时上报基金审核理事会，还组织赠款项目实施机构开展财务自查、并组织符合资质的会计师事务所对赠款项目实施机构开展财务检查和项目财务中期及其项目结项财务验收工作，以加强赠款项目资金管理、确保资金使用效果。

8.3.3　有偿使用项目资金管理

有偿使用项目资金是指基金管理中心按照《中国清洁发展机制基金管理办法》授权的使用方式，开展有利于产生应对气候变化效益的产业活动的资金。

基金管理中心按照财政部与国家发展和改革委员会联合批准的年度财务收支预算，开展基金有偿使用活动，并遵循会计核算的要求，及时、准确反映有偿使用项目资金状况、使用结果，监督有偿使用过程中的资产形成、处置和收益等情况。有偿使用项目实施机构需按照项目合同约定的内容、事项、时间、进度使用资金。有偿使用项目结束后，基金管理中心对项目资金使用执行情况、资金回收情况和项目绩效进行评估，并将评估结果分别报主管部门和基金审核理事会。

8.3.4　本外币现金理财活动

为充分发挥基金资金作用，实现基金资金的保值增值，基金管理中心对基金结余资金开展本外币现金理财活动。

基金管理中心以安全性、流动性和收益性，以及支持节能、能效提高、新能源和可再生能源等低碳活动为原则开展理财活动，具体包括结构性银行存款，购买国债、金融债和高信用等级的企业（公司）债券，以及从事风险等级较低的其他保值活动。基金资金的理财活动全部通过国内中资商业银行办理以控制风险。

通过基金的本外币现金理财活动，基金资金不仅可以发挥种子资金作用，撬动社会资金参与，而且可以盘活国内低碳发展资金，提高资金利用效率，更为重要的是还可以规范企业的温室气体减排统计方法，提升企业的减排能力。

第9章

中国清洁发展机制基金已开展的工作

自成立以来，基金管理中心以基金资金筹集、管理和使用为根本，紧紧围绕“打造四个平台，服务低碳发展”的宗旨准确定位（四个平台指资金平台、合作平台、行动平台和信息平台），从建章立制、国家收入收取、基金赠款项目管理、基金有偿使用项目开发与管理、重大问题研究、基金资金理财、服务企业低碳发展、加强国际合作以及公众意识提高等方面，全面开展工作，为国家应对气候变化和节能减排工作做好服务。

9.1 建章立制，确保基金的规范化运作

自基金筹备伊始，建章立制、能力建设就被列为基金管理中心的核心工作，以确保日后基金能够科学化、规范化、高标准运作。为此，基金管理中心编制了一系列机构行政管理、财务管理、人事管理、项目管理等方面的规章制度和工作规范，有效地保障了一个新建机构的安全、有序运行。

2010年9月14日，《中国清洁发展机制基金管理办法》正式颁布实施，明确界定了基金的性质、宗旨、使用原则、管理机构、基金资金的筹集与使用等。围绕《中国清洁发展机制基金管理办法》，在基金审核理事会的指导下，基金管理中心又制定了一系列细化的专项制度，包括《中国清洁发展机制基金审核理事会议事规则》、《中国清洁发展机制基金赠款项目管理办法》、《中国清洁发展机制基金有偿使用管理办法》、《中国清洁发展机制基金财务管理办法》、《中国清洁发展机制项目国家收入收取

办法》、《中国清洁发展机制基金委托贷款管理办法》、《中国清洁发展机制基金股权投资管理办法》、《中国清洁发展机制基金融资担保管理办法》等。目前，有关基金和基金管理中心业务开展和机构运行的规章制度已基本完善。

9.2 做好国家收入收取工作，为基金业务发展奠定基础

清洁发展机制项目转让温室气体减排量的国家收入是当前基金的最主要资金来源，故做好国家收入收取工作，是基金发展的根本保障。基金管理中心在国家收入收取方面已取得显著成绩。

9.2.1 规范业务流程，保证国家收入收取

基金管理中心针对国家收入收取工作环节多、流程长、对象复杂、工作任务繁重的特点，制定了《中国清洁发展机制项目国家收入收取实施细则》，细化和规范了内部、外部工作流程，设计开发了国家收入收取管理信息系统，保证了国家收入收取工作有条不紊地开展。

9.2.2 国家收入的收取情况

自2007年4月13日收到吉林洮南风力发电项目上缴的首笔国家收入人民币96,470元起，截至2011年6月30日，基金管理中心已累计收到来自370个项目的736批次核证减排量交易的国家收入，累计达80.42亿元人民币。

在上述国家收入中，因HFC-23分解类清洁发展机制项目实施较早，减排量规模大，且国家收入的收取比例为65%，该类项目产生的国家收入已达72.53亿元人民币，占国家收入总额的90.2%。目前我国该类项目已开发完毕，随着其他类型项目实施的逐步深入，该类项目在国家收入中占的比重将呈下降趋势。类似地，N_2O分解类清洁发展机制项目作为另一类工业废气类项目，虽然实施较晚，但因其减排量规模也普遍较大，国家收入的收取比

例为30%[1]，该类项目是国家收入来源的另一主力军，目前产生国家收入已达6.94亿元人民币，占国家收入总额的8.6%。而风电、水电、生物质发电等新能源和可再生能源，以及节能和能效提高等其他类型项目，虽然它们占已签发核证减排量项目数的95.3%、已签发核证减排量批次的83.5%，但因它们的减排量规模普遍较小，同时属于国家鼓励发展的项目，国家收入收取比例仅为2%，故它们产生的国家收入仅为0.96亿元，占国家收入总额的1.2%，如表9–1所示。

表 9–1　　我国各类清洁发展机制项目国家收入收取情况

行　业	征缴比例（%）	项目数	签发批次	到账金额（人民币万元）	比例（%）
HFC-23分解	65	11	104	725,308.60	90.2
N_2O分解	30	7	38	69,353.84	8.6
新能源和可再生能源	2	245	460	5,565.04	0.7
燃料替代	2	12	26	1,980.93	0.2
甲烷回收利用	2	13	31	1,356.92	0.2
节能和提高能效	2	33	65	671.68	0.1
合　计		489	1,125	804,237.01	100.0

注：以上数据截至2011年6月30日。

资料来源：中国清洁发展机制基金管理中心内部统计。

从上述国家收入的构成可以看到，工业废气类清洁发展机制项目对国家收入的贡献巨大。这印证了我国政府在应对气候变化问题上是有远见和负责任的，即通过设立中国清洁发展机制基金，调节和“绿化”工业废气类清洁发展机制项目企业的收入，消除了国际社会对我国这两类项目可能产生一些不良后果的顾虑，使清洁发展机制这一国际合作机制在我国应对气候变化、促进可持续发展方面，发挥更高层次、更全面、更长久的作用。

1　2011年8月3日颁布实施的《清洁发展机制项目运行管理办法》（修订）对N_2O分解类清洁发展机制项目的国家收入收取比例进行了调整，己二酸生产中的N_2O项目，国家收入收取比例为30%；硝酸等产生中的N_2O项目，国家收入收取比例为10%。

9.3　开展现金理财业务，确保资金安全和保值增值

基金管理中心自成立之初，就在严格资金管理，确保资金安全的同时，通过本外币银行存款，购买国债、金融债和高信用等级的企业（公司）债券，以及从事风险等级较低的其他保值活动等积极开展各种理财活动，努力利用现有授权开展理财，实现基金资金保值增值。截至2011年10月，基金管理中心通过各项理财活动，已实现运行收益约4亿元人民币。

9.3.1　外币理财业务

在基金国家收入中，美元和欧元等外币占国家收入总额的50%以上。而欧元因受金融危机影响，汇率大幅走低，人民币对美元汇率又持续走高，针对这一现状，基金管理中心密切跟踪国际主要经济体的宏观经济动向和国际外汇市场行情走势，对外币采取实时结汇与择机结汇相结合的方式，避免汇率波动风险造成损失。据初步统计，在人民币持续升值期间，基金管理中心避免汇率损失数亿元人民币。同时为开展外币理财业务做好前期铺垫和准备，积极利用汇率市场变化实现留存外币开展基金保值业务。基金管理中心通过中资银行，利用欧元头寸购买中短期外币理财产品、开展外汇掉期等外币理财业务，实现收益数百万元人民币。

9.3.2　本币理财业务

针对基金本币资金量大，存量资金充足的特点，基金管理中心通过开展银行定期理财、通知存款等无风险理财方式，最大限度地开展中长期定期理财业务，已取得现金理财业务增值数亿元的良好经济效益。

在实现基金资金保值增值的同时，基金管理中心还努力提升理财资金的社会化效益。基金管理中心积极与国内众多商业银行磋商，提出了“框架协议+执行协议”的合同合作模式，采取“1：N”杠杆模式，发挥种子资金作用，撬动社会资金共同开展节能减排活动。2010年和2011年基金管理中心

分别与浙商银行、兴业银行开展了清洁发展专项理财产品合作。在保证基金资金安全的同时，撬动了超过100%商业资金投入我国节能减排事业。据测算，该理财产品预计可实现二氧化碳减排60余万吨。这是基金管理中心首笔专项理财资金，也是基金管理中心将资金理财与应对气候变化和节能减排工作有机结合的大胆尝试和创新。其积极意义在于：创新机制，发挥市场功能，调动社会资金支持节能减排和应对气候变化事业；实现优势互补，降低基金资金运营风险，培育金融机构的低碳发展意识与减碳能力；积极探索，打造科学化精细化减排资金管理方式；资技互联，推动生产方式向低碳转变，支持发展方式的改革创新。

9.4 开展基金赠款管理，支持国家应对气候变化工作

应对气候变化工作已列入我国政府的重要议事日程。但是，我国在应对气候变化方面的基础工作、基本能力、基本认识还很薄弱，迫切需要加强，特别是在当今气候变化已成为国际社会几乎逢会必谈的重要议题的情况下，更急需迅速提高我国在这些方面的能力和意识，以适应对内、对外工作的需要。在此情况下，基金发挥自己独特的优势，通过赠款方式，有针对性地为国家急需的活动提供快速的资金支持。

9.4.1 基金赠款项目

2008年度，基金审核理事会批准了第一批15个基金赠款项目（见表9-2），总经费预算2,653万元。2011年基金出资2.6亿元资金全面支持我国应对气候变化与低碳发展工作，包括全国31个省市的应对气候变化方案制定及温室气体排放清单编制，五省八市（广东省、辽宁省、湖北省、陕西省、云南省，天津市、重庆市、深圳市、厦门市、杭州市、南昌市、贵阳市、保定市）低碳试点方案编制，五市两省（北京市、天津市、上海市、重庆市、深圳市、湖北省、广东省）碳交易试点方案的编制，以及相关部委应对气候变化的急需工作等120余个项目。

表 9-2　2008年度赠款项目清单

编号	项目名称	基金支持资金（万元）
1	清洁发展机制项目评审	650
2	《应对气候变化国家方案》更新	200
3	应对气候变化国家战略研究	300
4	主要缔约方在气候变化问题上的利益诉求、谈判立场及政策措施研究	150
5	电厂建设上大压小方法学开发	100
6	中国HFC-23排放控制战略研究	50
7	制作面向公众的气候变化宣传片	50
8	中国农业农村温室气体减排潜力评估研究	338
9	《气候变化中国在行动》对外宣传电视片	50
10	国家气候变化专家委员会咨询项目活动	50
11	后京都气候变化资金体系设计研究	60
12	应对气候变化财政政策研究	60
13	中国清洁发展机制基金机构能力建设	60
14	中国清洁发展机制基金制度机制建设	60
15	不同长期稳定浓度目标下气候情景、影响与应对策略研究	475
总计		2,653

9.4.2　赠款项目成果

已实施的基金赠款项目紧密围绕国家应对气候变化工作所需的国际谈判、政策决策支持、宣传教育和理论研究四大方面展开，从多角度多层面综合解决国家急需的科学支持、理论支持和舆论支持。这些赠款项目产生了重大影响。具体表现在：

第一，一些项目为国家制定我国应对气候变化政策、把气候变化工作纳入国家的整体发展规划，提供了重要的科学依据，比如更新的《应对气候变化国家方案》是我国应对气候变化领域非常重要的纲领性文件。另外一些项目，如不同长期稳定浓度目标下气候情景、影响与应对策略研究、中国农业农村温室气体减排潜力评估研究，则为我国提供了一些重要的基础研究成果，服务于国家的有关决策。

第二，一些项目，比如国家气候变化专家委员会咨询，主要缔约方在气候变化问题上的利益诉求、谈判立场及政策措施研究，应对气候变化国家战略研究等研究项目的产出，为我国更加积极、主动、有效地参与国际气候变化谈判提供了重要的理论支持和技术支持，维护了国家和发展中国家的利益。

第三，一些项目活动和成果，比如制作面向公众的气候变化宣传片和《气候变化中国在行动》对外宣传电视片两项目的产出，在国际和国内很好地宣传了我国在应对气候变化工作中所做的努力，增进了国际社会对我国的了解，提升了我国公众对气候变化问题的认识，产生了积极的国际影响和社会影响。

第四，一些项目有力地支持了我国的清洁发展机制项目开展，比如清洁发展机制项目评审等，使得清洁发展机制项目在我国各地迅速开展，并很快在清洁发展机制项目的国家批准项目数、联合国注册项目数和核证减排量签发量这三项重要指标上在全球排名第一，使我国成为全球碳市场最重要的组成部分之一。

这些项目切入点准确、立意深远、批准迅速，为我国政府制定有关应对气候变化政策和战略，参加或组织2008年、2009年、2010年众多国际国内气候变化会议制定谈判对策，开展国内外宣传，提供了及时、有力的支持，充分体现了基金的独特而重要的作用。

9.5 积极推进基金有偿使用工作，支持节能减排

国务院已明确基金是“按社会性基金模式管理的政策性基金”，这是对基金运作管理方式的基本定位。创新机制、推动节能减排和应对气候变化事业的科学化、市场化、专业化，是基金的根本目标和任务。作为政策性基金，它将体现政府政策导向。按照社会性基金模式管理，意味着它可以进入广阔市场，同众多商业机构开展灵活合作，进行市场运作。两者的结合将架

起直接连接政府政策和市场发展的桥梁。基金管理中心自成立伊始，就对基金有偿使用工作的战略、重点领域、实施手段等进行了全面探索，为基金以有偿使用方式支持国家节能减排政策目标做好准备。

9.5.1 明确基金有偿使用的战略指导思想

目前，基金资金规模已达80亿元人民币，预计到2012年年底可达122亿元人民币。这些资金与每年国家投入到应对气候变化和节能减排的财政资金相比，可谓杯水车薪，与金融机构和企业投资于相关领域的资金量相比，更是微乎其微。但基金作为国家财政的一种创新机制，应利用新机制、新机构和新成员的特点，在国内发展低碳经济、应对气候变化事业中应发挥创新、先锋、引导和示范作用。

基金管理中心通过走访大量国内金融机构、商业银行、地方政府和企业，参照世界银行、亚洲开发银行、社保基金、中比基金，以及英国碳基金等采用的项目运作模式等，把应对气候变化事业产业化、市场化和社会化作为基金长期奋斗目标和任务，解决了基金发展动力、发展手段和社会广泛参与等问题，并提出了“借船下海、专业运作、捆绑发展”的战略指导思想，即以基金的有限资金作为种子资金，撬动更多的市场和社会资源参与节能减排，把科学发展、创新发展和精细发展的要求落到实处。同时，这也有效解决了当前机构自身能力偏弱，控制市场风险、行业机构道德风险能力较差等问题，弥补了自身的短板。

针对当前我国温室气体排放主要集中在工业，大量工业生产技术含量低，工艺、管理落后，节能减排潜力巨大的现实，基金管理中心提出了基金有偿使用工作的“市场减排、技术减排和社会减排”三个着力点。这三个着力点是基金“三化”长期目标下现阶段工作的具体化，是一个行动纲领。

考虑到基金是政策性基金，而非商业金融机构，不追求经济利益最大化的特点，基金管理中心特别重视基金有偿资金使用的社会效益。在有偿使用项目筛选过程中，除考虑经济效益和风险外，把碳减排效果作为重要衡量指标之一，要求所有申报项目均需提交碳减排预算报告；在有偿使用项目实施过程中，所有项目均需建立碳资产管理台账，核算项目的温室气体减排情况。这既支持了国家的节能减排工作，又提高了受资助企业的碳资产管理意识，体现基金的政策引导性作用。

9.5.2 探索有偿资金使用方式

在委托贷款方面，基金管理中心探索通过地方政府部门、商业银行、国有大型企业三种渠道，向符合条件的企业提供一定期限且利率低于同期中国人民银行指导利率的优惠清洁发展委托贷款。基金率先启动了与地方政府部门开展清洁发展委托贷款的合作模式，截至2011年10月，已在陕西、山西、河北、湖南、福建、江苏、山东以及江西等8个省份开展了21个清洁发展委托贷款项目，贷款规模超过10亿元人民币，预计项目投产后每年可减少温室气体排放约1,000万吨二氧化碳当量。

在股权投资方面，基金管理中心在符合国家政策导向的前提下，根据资本市场运作特点，结合基金现状，借助金融机构的客户资源和经验优势，筛选具有高成长性的企业开展股权投资合作。

在融资性担保方面，基金管理中心正加紧与专业担保机构共同探索开展融资性担保业务的模式，并与一些担保公司共同研究和商谈担保合作事宜，以期在全面开展基金有偿使用业务的同时，提高和完善机构的项目运作能力。

清洁发展委托贷款项目案例

福建省X市C公司风电场项目

福建省A公司属大型国有独资公司，注册资本为59亿元。截至2010年年底，公司总资产389亿元，其中所有者权益224亿元。2008年12月，A公司（股权占51%）联合福建省B公司（股权占49%）共同出资8,000万元设立C公司，主要从事风力发电项目的投资。

为合理开发和充分利用沿海地区丰富的风能资源，缓解对一次能源的依赖，减轻环境污染，也为A公司未来大规模兴建风力发电项目做好示范，该公司计划投资6亿元，在X市开发建设风力发电项目，拟安装24台荷兰维斯塔斯2MW的永磁直趋风力发电机组，以并网形成48MW的风力发电规模，同时建设1座110kV升压变电站。在完成现场测风实验后，该项目已于2011年2月动工建设，预计年底投产。为缓解建设资金压力，福建省投资集团通过福建省财政厅向我基金申请清洁发展委托贷款7,000万元，占企业总投资额的12%。

（1）社会效益。包括风能在内的新能源开发利用，符合国家产业政策及国民经济可持续发展的要求。充分利用国内丰富的风能资源，有利于减少对一次能源的依赖，实现能源资源的合理开发利用和优化配置，促进能源与环境的协调发展，在增加当地居民就业机会的同时，改善我国的能源结构。

（2）环境效益。本项目符合国家能源建设方向和福建省能源电力发展规划。据估算，项目建成后，在每年可向电网输电138GW·h的同时，能够减碳11.23万吨，并可减少多种大气污染物和大量灰渣的排放，有利于大气质量的改善。

（3）经济效益。项目投产后，可实现年销售收入7,195万元，净利润1,216.35万元。

鉴于风电场投资具有回收期长且初期投资较大等特点，为了降低投资风险，C公司提出由其母公司——A公司负责偿还贷款。A公司2010年实现净利润7.32亿元。该公司未来三年的整体运营状况表明该公司可满足按期还款的需要。

经过基金管理中心内部评审程序，通过风险管理委员会和投资评审委员会评审，并报财政部及国家发展和改革委员会备案，基金管理中心最终批复对C公司贷以3年期，利率5.6525%，金额5,000万元人民币的款项加以支持，撬动社会资金达我资金的11倍。

9.6 开展政策研究，发挥智库作用

基金管理中心自成立以来，密切关注气候变化国际合作进展、国际碳交易市场发展和国内气候变化和低碳发展的行动动态，结合基金业务发展需

要，组织开展了系列政策研究，为我国应对气候变化和低碳发展的政策制定与行动开展中更好发挥基金的创新机制作用提出建议。

9.6.1 开展气候变化融资问题研究和市场化减排机制研究

资金问题是应对气候变化的一个关键问题，也是气候变化国际合作的关键问题。基金管理中心利用在基金筹备和业务开展过程中的政策研究积累，编写出版了我国第一本比较全面地介绍气候变化融资问题的专著《气候变化融资》。基金管理中心还根据基金工作定位，为促进节能减排资金的科学化、精细化管理提出建议。

基金管理中心密切跟踪国际上利用市场机制促进应对气候变化和低碳发展的行动进展，包括全球碳交易市场和我国碳交易的发展状况，将最新信息、分析报告及时提交政府部门参考或在期刊、报纸、网站上刊登介绍。

9.6.2 开展“三可”研究，推进国内碳市场“三大平台”建设

为加强体现基金业务的政策性引导作用，提高基金有偿使用业务的附加价值，并为逐步建立国内碳交易体系的基础设施建设进行探索，基金管理中心根据温室气体减排“可测量、可报告、可核查”（以下简称“三可”）的工作需要，组织开展了针对性的政策研究，并与行业机构合作，对若干高耗能行业开展碳减排标准的起草进行示范推动，引导这些行业开展碳减排标准的制定工作，此外还筹建了核准认证平台，启动了基金有偿使用项目碳预算编制和认证评估工作。基金管理中心为国内碳市场建立开展的准备工作，涵盖了碳市场建设的认证平台、交易平台和清算平台等各个环节。

9.7 国际合作

基金是应对气候变化国际合作的一个产物，基金管理中心积极发挥国际合作优势，将其他国家的先进观念、先进技术、实践经验等引入国内，

开展机构能力建设、信息和知识共享等务实合作，促进我国的应对气候变化工作。

基金管理中心多次参加联合国气候变化大会、碳博览会（Carbon Expo）、亚洲碳论坛（Carbon Asia Forum）、亚洲开发银行清洁能源周等活动，通过举办展览和主题分会，在大会和分会上演讲，接受国外媒体采访等方式，向国际社会介绍基金及其在中国应对气候变化、低碳发展中的作用，提高了基金在全球业界的知名度和影响力。

9.8　低碳发展公众意识宣传

应对气候变化和低碳发展需要广泛的公众参与，提高公众意识至关重要，基金将此作为一项重要的工作任务。基金管理中心自成立以来，利用建立基金网站和发行内部刊物、参加国内外重要会议、接受媒体采访、开展业务交流和赠款项目支持等方式，积极开展宣传。

基金管理中心已开设了自己的官方网站（http://www.cdmfund.org），该网站已成为我国专门开展应对气候变化工作宣传的重要网站之一。基金管理中心还经常通过报刊、电视等媒体向社会公众介绍应对气候变化和低碳发展的国际国内最新动态，以及基金最新工作进展。

近年来，基金管理中心连续召开全国财政系统应对气候变化与低碳发展培训研讨会议，促进地方财政部门提高有关应对气候变化与低碳发展的认识并积极参与其中，同时为基金管理中心与地方财政部门的合作建立基础。

第10章

中国清洁发展机制基金的展望

在社会经济发展的重要战略转型期，我国政府提出了“五个坚持”，要求在未来五年的改革中全面强化“转变经济发展方式”，以科学发展为主线，推进社会和谐可持续发展。气候变化影响社会经济发展的环境和资源空间。作为国家应对气候变化的创新资金机制，基金将在国家统一部署下，充分发挥并深化平台作用，积极引导、支持和推动开展全面、系统的应对气候变化行动，促进应对气候变化和低碳发展的产业化、市场化和社会化。

第一，要保证基金运行平稳、安全，这是基金发展的根本所在。基金管理中心将在继续做好国家收入收取的同时，一方面努力开拓基金资金的新来源，另一方面积极探索新的金融理财模式，提高资金运营效率，严把风险控制，实现基金资金保值增值。

第二，创新、支持、引导、催化是基金业务的主旋律。在赠款方面，基金将配合财政主渠道，合理安排赠款资金，积极支持国家和地方开展应对气候变化政策研究、能力建设和公众意识提高等工作。在有偿使用方面，基金将重点关注清洁能源、可再生能源和工业节能等领域，通过投资、贷款、担保等方式，支持减排效益大、经济效益好、示范作用明显的项目，并撬动更多的社会资源参与应对气候变化和低碳发展。

第三，配合国家节能能效审计工作。基金将在“十二五”期间，以推动市场机制减排、技术减排和新兴产业减排为目标，利用独特平台优势，联合国内外权威研究机构，探索我国碳减排的政策和市场机制创新，推动国内碳排放交易市场建立。基金将在有偿使用项目中实施减碳预算编制和碳盘查，通过项目，引导、提升公众的碳资产管理意识和能力，最终推动国家建立碳减排量和节能量登记、核证和交易平台体系，为国家碳减排标准化、透明化、可量化工作做好先行试点。

第四，在应对气候变化技术促进方面，基金将以多种形式支持重点行业、关键领域、关键环节的突破性技术的自主研发，鼓励相关技术和措施的研究开发和应用普及。此外，基金还将充分发挥信息与交流平台作用，示范推广先进技术和理念，并通过信息的系统收集和整理、国际交流、能力培训等活动，实现知识、信息和技术的交流和共享。

在“十二五”规划的宏伟蓝图下，基金将大有可为且任重道远。基金将把握机遇，开拓创新，力求实效，与国内和国际各界一起，为中国和全球应对气候变化和低碳发展事业发挥自己独特、重要的作用。

中国清洁发展机制基金大事记

1. 2005年10月，中国政府启动中国清洁发展机制基金筹备工作。

2. 2005年12月，根据与中国政府达成的协议，世界银行提前支付其购买的、2个中国HFC-23分解清洁发展机制项目产生的第一批核证减排量对应的部分国家收入630万美元，用于基金筹建和初期运行。

3. 2006年5月，亚洲开发银行为基金提供60万美元技术援助项目，用于基金的筹建和机构能力加强。

4. 2006年8月，国务院批准建立中国清洁发展机制基金及其管理中心。

5. 2007年3月，基金管理中心在国家事业单位登记管理局完成注册登记。

6. 2007年4月，基金审核理事会召开第一次会议，确定国家发展和改革委为基金主席，财政部为副主席，并正式启动基金相关制度的编制工作。

7. 2007年4月13日，基金正式收到第一笔清洁发展机制项目国家收入。

8. 2007年6月，中国政府发布《应对气候变化国家方案》，其中指出要“有效利用中国清洁发展机制基金”。

9. 2007年11月9日，财政部和国家发改委联合启动了基金的业务运行。

10. 2008年10月，亚洲开发银行提供80万美元的技术援助项目，用于帮助基金初期的机构能力建设和初期业务运行。

11. 2008年11月，国家发展改革委、财政部联合批准第一批清洁基金赠款项目，支持急需的国家应对气候变化工作。

12. 2009年3月，财政部颁布《关于中国清洁发展机制基金及清洁发展机制项目实施企业有关所得税政策问题的通知》，为基金和相关清洁发展机制项目企业减轻了税负。

13. 2009年4月，基金战略发展委员会成立并召开第一次会议。

14. 2010年9月14日，经国务院批准，财政部、国家发改委等七部委联合颁布了《中国清洁发展机制基金管理办法》。

15. 2010年11月，基金与浙商银行合作，开辟现金理财新模式——清洁发展专项理财。

16. 2011年4月，基金第一批清洁发展委托贷款项目获得批准，标志着基金有偿使用业务实质启动。

17. 2011年12月7日，基金与陕西省人民政府战略合作协议签署仪式举行，标志着基金支持地方绿色低碳发展工作开始从单个项目的点上升到省级区域的面。

18. 2011年12月1日，《"十二五"控制温室气体排放工作方案》发布，明确要充分利用中国清洁发展机制基金资金，拓宽多元化投融资渠道，积极引导社会资金、外资投入低碳技术研发、低碳产业发展和控制温室气体排放重点工程。

19. 2011年12月23日，以基金为并列第一大股东的上海环境能源交易所股份有限公司揭牌仪式成功举行，标志着基金股权投资业务的开始，也标志着基金着手参与碳市场战略布局，助力国内碳市场建设。

20. 2011年12月31日，作为基金主要资金来源的清洁发展机制项目国家收入达到100亿元人民币。

附　录

附　录 1

中华人民共和国财政部
中华人民共和国国家发展和改革委员会
中华人民共和国外交部
中华人民共和国科学技术部　　令
中华人民共和国环境保护部
中华人民共和国农业部
中　国　气　象　局

第59号

中国清洁发展机制基金管理办法

第一章　总则

第一条　为加强和规范中国清洁发展机制基金（以下简称基金）的资金筹集、管理和使用，实现基金宗旨，制定本办法。

第二条　基金是由国家批准设立的按照社会性基金模式管理的政策性基金。

第三条　基金的宗旨是支持国家应对气候变化工作，促进经济社会可持续发展。

第四条　基金的筹集、管理和使用，应当遵循公开、公正、安全、效率、专款专用的原则。

第二章　管理机构及其职责

第五条　基金的管理机构由基金审核理事会和基金管理中心组成。

第六条　基金审核理事会是关于基金事务的部际议事机构。

基金审核理事会由国家发展改革委、财政部、外交部、科技部、环境保护部、农业部和气象局的代表组成。

基金审核理事会设主席和副主席，分别由国家发展改革委和财政部派出代表履行职责。

基金审核理事会负责审核下列事项：

（一）基金基本管理制度；

（二）基金发展战略规划，包括资金使用年度计划；

（三）基金赠款项目和重大有偿使用项目申请；

（四）基金年度财务收支预算与决算；

（五）基金其他重大业务事项。

前款所列事项经基金审核理事会审核并取得一致意见后，报国家发展改革委、财政部批准。

第七条　基金管理中心是基金的日常管理机构，具体负责基金的筹集、管理和使用工作，由财政部归口管理。

第八条　基金管理中心履行下列职责：

（一）起草基金基本管理制度，制定基金具体运行管理规定；

（二）筹集基金资金；

（三）管理基金资金，组织开展基金的有偿使用和理财活动；

（四）编制并组织实施基金的年度财务收支预算与决算；

（五）监督管理基金所支持项目的运行；

（六）向基金审核理事会报告基金的重大业务事项；

（七）开展其他符合基金宗旨的活动。

第三章　基金筹集

第九条　基金来源包括：

（一）通过清洁发展机制项目转让温室气体减排量所获得收入中属于国家所有的部分；

（二）基金运营收入；

（三）国内外机构、组织和个人捐赠；

（四）其他来源。

第十条 本办法所称减排量，是指经国家批准，通过清洁发展机制项目转让的温室气体减排量；减排量收入，是指转让减排量所获得的收入。

减排量收入由国家和实施清洁发展机制项目的企业（以下称项目业主）按照规定的比例分别所有。减排量收入中属于国家所有的部分（以下称国家收入）全额纳入基金。

第十一条 国家收入由基金管理中心负责向项目业主或按减排量转让合同约定向减排量购买方收取。

项目业主应当在取得减排量收入后的15个工作日内，按照规定比例向指定账户支付国家收入。

第十二条 国家收入应当以减排量转让合同约定的币种取得。

减排量转让合同约定以外币支付，但确需以人民币支付国家收入的，经基金管理中心同意，项目业主应当在取得收入后的15个工作日内以人民币支付，汇率以结汇日现汇买入价为准。

第十三条 减排量转让合同由项目业主和减排量购买方签订。

项目业主应当在减排量转让合同生效后15个工作日内，将合同副本、营业执照复印件、合同双方联系人及联系方式报基金管理中心备案。

备案事项发生变更的，项目业主应当自变更之日起15个工作日内告知基金管理中心。

第十四条 项目业主缓缴、少缴、不缴国家收入的，由财政部、国家发展改革委依据有关规定予以处理、处罚。

第四章 基金使用

第十五条 基金使用采取赠款、有偿使用等方式。

基金通过赠款方式支持有利于加强应对气候变化能力建设和提高公众应对气候变化意识的相关活动。

基金通过有偿使用方式支持有利于产生应对气候变化效益的产业活动。

基金通过银行存款、购买国债、金融债、企业债等形式开展理财活动。

第十六条 基金支出包括业务支出和基础管理费支出。

业务支出包括赠款支出和有偿使用项目开发费用支出。

基金赠款年度支出规模根据国家应对气候变化实际工作需要确定。

本办法所称有偿使用项目开发费用，是指基金有偿使用项目筛选、调查、评审、立项过程中发生的费用。有偿使用项目开发费用按照项目使用金额的一定比例提取。

本办法所称基础管理费支出，是指基金筹集、管理、使用过程中的日常管理费用，包括清洁发展机制项目日常管理费用。基础管理费支出按照基金上年末资产净值的一定比例提取。

有偿使用项目开发费用和基础管理费支出的具体提取比例由基金审核理事会另行规定。

第十七条 基金管理中心应当对基金使用进行风险控制。

基金不得用于不符合其宗旨的赞助和捐赠支出，不得从事股票、股票类投资基金、房地产以及期货等金融衍生产品投资。

第十八条 基金与基金管理中心财务应当分别建账，分别核算，实行预决算管理。

财政部负责制定基金财务管理办法，并对基金使用情况和会计记录进行监督检查。

第五章 赠款项目管理

第十九条 赠款主要用于支持下列事项：

（一）与应对气候变化相关的政策研究和学术活动；

（二）与应对气候变化相关的国际合作活动；

（三）旨在加强应对气候变化能力建设的培训活动；

（四）旨在提高公众应对气候变化意识的宣传、教育活动；

（五）服务于基金宗旨的其他事项。

第二十条 赠款项目申请人应当是我国境内从事应对气候变化领域工作，具有一定研究或者培训能力的相关机构。

第二十一条 申请赠款应当提交项目申请书。赠款项目申请书包括以下内容：

（一）申请人基本情况；

（二）项目背景资料；

（三）项目目标；

（四）项目的主要内容与活动；

（五）项目的主要产出；

（六）项目的执行进度安排；

（七）申请资金额和预算安排；

（八）其他相关内容。

第二十二条 赠款项目申请书由国务院有关部门或者省级发展改革部门（以下称项目组织申报单位）向国家发展改革委转报或报送。

第二十三条 国家发展改革委负责组织赠款项目的评审。

赠款项目的评审结果报基金审核理事会审核并取得一致意见后，由国家发展改革委、财政部批准。

第二十四条 赠款项目由项目组织申报单位组织实施。

第二十五条 赠款项目实行合同管理，在合同中明确规定各方责任、权利、义务和违约处罚办法。

赠款项目合同由国家发展改革委、项目组织申报单位、基金管理中心、赠款项目申请人共同签订。

第二十六条 国家发展改革委、基金管理中心会同项目组织申报单位负责对赠款项目的实施进行监督检查和考核验收。国家发展改革委、财政部对违规行为予以处理、处罚。

第二十七条 赠款项目形成研究或者其他成果的，有关权益归属在赠款项目合同中约定。

第六章　有偿使用项目管理

第二十八条 基金有偿使用采取以下方式：

（一）股权投资；

（二）委托贷款；

（三）融资性担保；

（四）国家批准的其他方式。

基金以股权投资、委托贷款方式支持项目的，其年度累积金额不得超过上年末资产净值的一定比例。具体比例由基金审核理事会另行规定。

基金以股权投资方式支持项目的，不得对投资对象控股，投资所形成股权的退出，应当按照公开、公平和市场化原则，确定退出方式及退出价格。

基金以融资担保方式支持项目的，其担保额不得超过基金年度预算确定的限额。

第二十九条 有偿使用项目申请人应当是我国境内从事减缓、适应气候变化相关领域业务的中资企业、中资控股企业。

第三十条　有偿使用项目申请人应当向基金管理中心提交申请文件。申请文件包括以下内容：

（一）项目申请书；

（二）项目可行性研究报告；

（三）企业近3年经营状况；

（四）企业营业执照副本；

（五）其他相关材料。

第三十一条　基金管理中心负责组织对基金有偿使用项目的遴选、评审。

属于重大项目的，应当报经基金审核理事会审核并取得一致意见后，由国家发展改革委、财政部批准；属于非重大项目的，由基金管理中心按照规定程序审批，并于批准后的15个工作日内报国家发展改革委、财政部备案。

前款所称重大项目是指单个项目申请基金资金在7,000万元人民币以上（含7,000万元）的有偿使用项目。

第三十二条　按照国家有关投资管理规定，应当办理项目审批、核准或者备案手续的，从其规定。

在项目未获得审批、核准或者备案前，基金不得为项目提供资金。

第三十三条　基金管理中心负责有偿使用项目的组织实施、监督检查和考核验收。

基金有偿使用形成的各种资产及权益应当按照国家有关财务规章制度进行管理。

第七章　附则

第三十四条　基金及其管理中心应当接受国家审计机关依法实施的审计监督。

第三十五条　经基金审核理事会批准，基金管理中心可以聘请社会审计机构对基金收支规模、基金结余、基金运行情况以及基金管理中心的支出情况进行审计。

第三十六条　本办法自发布之日起施行。

附　录2

中华人民共和国国家发展和改革委员会
中华人民共和国外交部
中华人民共和国科学技术部
中华人民共和国财政部　令

第11号

清洁发展机制项目运行管理办法

（修订）

第一章　总则

第一条　为促进和规范清洁发展机制项目的有效有序运行，履行《联合国气候变化框架公约》（以下简称《公约》）、《京都议定书》（以下简称《议定书》）以及缔约方会议的有关决定，根据《中华人民共和国行政许可法》等有关规定，制定本办法。

第二条　清洁发展机制是发达国家缔约方为实现其温室气体减排义务与发展中国家缔约方进行项目合作的机制，通过项目合作，促进《公约》最终目标的实现，并协助发展中国家缔约方实现可持续发展，协助发达国家缔约方实现其量化限制和减少温室气体排放的承诺。

第三条　在中国开展清洁发展机制项目应符合中国的法律法规，符合《公约》、《议定书》及缔约方会议的有关决定，符合中国可持续发展战略、政策，以及国民经济和社会发展的总体要求。

第四条　清洁发展机制项目合作应促进环境友好技术转让，在中国开展合作的重点领域为节约能源和提高能源效率、开发利用新能源和可再生能源、回收利用甲烷。

第五条　清洁发展机制项目的实施应保证透明、高效，明确各项目参与方的责任与义务。

第六条　在开展清洁发展机制项目合作过程中，中国政府和企业不承担《公约》和《议定书》规定之外的任何义务。

第七条　清洁发展机制项目国外合作方用于购买清洁发展机制项目减排量的资金，应额外于现有的官方发展援助资金和其在《公约》下承担的资金义务。

第二章　管理体制

第八条　国家设立清洁发展机制项目审核理事会（以下简称项目审核理事会）。项目审核理事会组长单位为国家发展改革委和科学技术部，副组长单位为外交部，成员单位为财政部、环境保护部、农业部和中国气象局。

第九条　国家发展改革委是中国清洁发展机制项目合作的主管机构，在中国开展清洁发展机制合作项目须经国家发展改革委批准。

第十条　中国境内的中资、中资控股企业作为项目实施机构，可以依法对外开展清洁发展机制项目合作。

第十一条　项目审核理事会主要履行以下职责：

（一）对申报的清洁发展机制项目进行审核，提出审核意见；

（二）向国家应对气候变化领导小组报告清洁发展机制项目执行情况和实施过程中的问题及建议，提出涉及国家清洁发展机制项目运行规则的建议。

第十二条　国家发展改革委主要履行以下职责：

（一）组织受理清洁发展机制项目的申请；

（二）依据项目审核理事会的审核意见，会同科学技术部和外交部批准清洁发展机制项目；

（三）出具清洁发展机制项目批准函；

（四）组织对清洁发展机制项目实施监督管理；

（五）处理其他相关事务。

第十三条　项目实施机构主要履行以下义务：

（一）承担清洁发展机制项目减排量交易的对外谈判，并签订购买协议；

（二）负责清洁发展机制项目的工程建设；

（三）按照《公约》、《议定书》和有关缔约方会议的决定，以及与国外合作方签订购买协议的要求，实施清洁发展机制项目，履行相关义务，并接受国家发展改革委及项目所在地发展改革委的监督；

（四）按照国际规则接受对项目合格性和项目减排量的核实，提供必要的资料和监测记录。在接受核实和提供信息过程中依法保护国家秘密和商业秘密；

（五）向国家发展改革委报告清洁发展机制项目温室气体减排量的转让情况；

（六）协助国家发展改革委及项目所在地发展改革委就有关问题开展调查，并接受质询；

（七）企业资质发生变更后主动申报；

（八）根据本办法第三十六条规定的比例，按时足额缴纳减排量转让交易额；

（九）承担依法应由其履行的其他义务。

第三章　申请和实施程序

第十四条　附件所列中央企业直接向国家发展改革委提出清洁发展机制合作项目的申请，其余项目实施机构向项目所在地省级发展改革委提出清洁发展机制项目申请。有关部门和地方政府可以组织企业提出清洁发展机制项目申请。国家发展改革委可根据实际需要适时对附件所列中央企业名单进行调整。

第十五条　项目实施机构向国家发展改革委或项目所在地省级发展改革委提出清洁发展机制项目申请时必须提交以下材料：

（一）清洁发展机制项目申请表；

（二）企业资质状况证明文件复印件；

（三）工程项目可行性研究报告批复（或核准文件，或备案证明）复印件；

（四）环境影响评价报告（或登记表）批复复印件；

（五）项目设计文件；

（六）工程项目概况和筹资情况说明；

（七）国家发展改革委认为有必要提供的其他材料。

第十六条　如果项目在申报时尚未确定国外买方，项目实施机构在填报项目申请表时必须注明该清洁发展机制合作项目为单边项目。获国家批准后，项目产生的减排量将转入中国国家账户，经国家发展改革委批准后方可将这些减排量从中国国家账户中转出。

第十七条　国家发展改革委在接到附件所列中央企业申请后，对申请材料不齐全或不符合法定形式的申请，应当场或在五日内一次告知申请人需要补正的全部内容。

第十八条　项目所在地省级发展改革委在受理除附件所列中央企业外的项目实施机构申请后二十个工作日内，将全部项目申请材料及初审意见报送国家发展改革委，且不得以任何理由对项目实施机构的申请作出否定决定。对申请材料不齐全或不符合法定形式的申请，项目所在地省级发展改革委应当场或在五日内一次告知申请人需要补正的全部内容。

第十九条　国家发展改革委在受理本办法附件所列中央企业提交的项目申请，或项目所在地省级发展改革委转报的项目申请后，组织专家对申请项目进行评审，评审时间不超过三十日。项目经专家评审后，由国家发展改革委提交项目审核理事会审核。

第二十条　项目审核理事会召开会议对国家发展改革委提交的项目进行审核，提出审核意见。项目审核理事会审核的内容主要包括：

（一）项目参与方的参与资格；

（二）本办法第十五条规定提交的相关批复；

（三）方法学应用；

（四）温室气体减排量计算；

（五）可转让温室气体减排量的价格；

（六）减排量购买资金的额外性；

（七）技术转让情况；

（八）预计减排量的转让期限；

（九）监测计划；

（十）预计促进可持续发展的效果。

第二十一条　国家发展改革委根据项目审核理事会的意见，会同科学技术部和外交部作出是否出具批准函的决定。对项目审核理事会审核同意批准的项目，从项目受理之日起二十个工作日内（不含专家评审的时间）办理批准手续；对项目审核理事会审核同意批准，但需要修改完善的项目，在接到项目实施机构提交的修改完善材料后会同科学技术部和外交部办理批准手续；对项目审核理事会审核不同意批准的项目，不予办理批准手续。

第二十二条　项目经国家发展改革委批准后，由经营实体提交清洁发展机制执行理事会申请注册。

第二十三条 国家发展改革委负责对清洁发展机制项目的实施进行监督。项目实施机构在清洁发展机制项目成功注册后十个工作日内向国家发展改革委报告注册状况，在项目每次减排量签发和转让后十个工作日内向国家发展改革委报告签发和转让有关情况。

第二十四条 工程建设项目的审批程序和审批权限，按国家有关规定办理。

第四章 法律责任

第二十五条 本办法涉及的行政机关及其工作人员，在清洁发展机制项目申请过程中，对符合法定条件的项目申请不予受理，或当项目实施机构提交的申请材料不齐全、不符合法定形式时，不一次告知项目实施机构必须补正的全部内容的，由其上级行政机关或者监察机关责令改正；情节严重的，对直接负责的主管人员和其他直接责任人员依法给予行政处分。

第二十六条 本办法涉及的行政机关及其工作人员，在接收、受理、审批项目申请，以及对项目实施监督检查过程中，索取或者收受他人财物或者谋取其他利益，构成犯罪的，依法追究刑事责任；尚不构成犯罪的，依法给予行政处分。

第二十七条 本办法涉及的行政机关及其工作人员，对不符合法定条件的项目申请予以批准，或者超越法定职权作出批准决定的，由其上级行政机关或者监察机关责令改正，对直接负责的主管人员和其他直接责任人员依法给予行政处分；构成犯罪的，依法追究刑事责任。

第二十八条 项目实施机构在清洁发展机制项目申请及实施过程中，如隐瞒有关情况或者提供虚假材料的，国家发展改革委可不予受理或者不予行政许可，并给予警告。

第二十九条 项目实施机构以欺骗、贿赂等不正当手段取得批准函的，国家发展改革委依法处以与项目减排量转让收入相当的罚款，罚款收入按照《行政处罚法》等有关规定，就地上缴中央国库。构成犯罪的，依法追究刑事责任。

第三十条 项目实施机构在取得国家发展改革委出具的批准函后，企业股权变更为外资或外资控股的，自动丧失清洁发展机制项目实施资格，股权变更后取得的项目减排量转让收入归国家所有。

第三十一条　项目实施机构在减排量交易完成后，未按照相关规定向国家按时足额缴纳减排量交易额分成的，国家发展改革委依法对项目实施机构给予行政处罚。

第三十二条　项目实施机构伪造、涂改批准函，或在接受监督检查时隐瞒有关情况、提供虚假材料或拒绝提供相关材料的，国家发展改革委依法给予行政处罚；构成犯罪的，依法追究刑事责任。

第五章　附则

第三十三条　本办法中的发达国家缔约方是指《公约》附件一中所列的国家。

第三十四条　本办法中的清洁发展机制执行理事会是指《议定书》下为实施清洁发展机制项目而专门设置的管理机构。

第三十五条　本办法中的经营实体是指由清洁发展机制执行理事会指定的审定和核证机构。

第三十六条　清洁发展机制项目因转让温室气体减排量所获得的收益归国家和项目实施机构所有，其他机构和个人不得参与减排量转让交易额的分成。国家与项目实施机构减排量转让交易额分配比例如下：

（一）氢氟碳化物（HFC）类项目，国家收取温室气体减排量转让交易额的65%；

（二）己二酸生产中的氧化亚氮（N_2O）项目，国家收取温室气体减排量转让交易额的30%；

（三）硝酸等生产中的氧化亚氮（N_2O）项目，国家收取温室气体减排量转让交易额的10%；

（四）全氟碳化物（PFC）类项目，国家收取温室气体减排量转让交易额的5%；

（五）其他类型项目，国家收取温室气体减排量转让交易额的2%。

国家从清洁发展机制项目减排量转让交易额收取的资金，用于支持与应对气候变化相关的活动，由中国清洁发展机制基金管理中心根据《中国清洁发展机制基金管理办法》收取。

第三十七条　国家发展改革委已批准项目2012 年后产生的减排量，须经国家发展改革委同意后才可转让，项目实施按照本办法管理。

第三十八条　本办法由国家发展改革委商科学技术部、外交部、财政部解释。

第三十九条 本办法自发布之日起施行。2005 年10月12日起实施的《清洁发展机制项目运行管理办法》即行废止。

附：

可直接向国家发展改革委提交清洁发展机制项目申请的中央企业名单

1. 中国核工业集团公司
2. 中国核工业建设集团公司
3. 中国化工集团公司
4. 中国化学工程集团公司
5. 中国轻工集团公司
6. 中国盐业总公司
7. 中国中材集团公司
8. 中国建筑材料集团公司
9. 中国电子科技集团公司
10. 中国有色矿业集团有限公司
11. 中国石油天然气集团公司
12. 中国石油化工集团公司
13. 中国海洋石油总公司
14. 国家电网公司
15. 中国华能集团公司
16. 中国大唐集团公司
17. 中国华电集团公司
18. 中国国电集团公司
19. 中国电力投资集团公司
20. 中国铁路工程总公司
21. 中国铁道建筑总公司
22. 神华集团有限责任公司
23. 中国交通建设集团有限公司
24. 中国农业发展集团总公司
25. 中国林业集团公司
26. 中国铝业公司

27. 中国航空集团公司
28. 中国中化集团公司
29. 中粮集团有限公司
30. 中国五矿集团公司
31. 中国建筑工程总公司
32. 中国水利水电建设集团公司
33. 国家核电技术有限公司
34. 中国节能投资公司
35. 中国中煤能源集团公司
36. 中国煤炭科工集团有限公司
37. 中国机械工业集团有限公司
38. 中国中钢集团公司
39. 中国冶金科工集团有限公司
40. 中国钢研科技集团公司
41. 中国广东核电集团

参考文献

［1］政府间气候变化专门委员会. 气候变化2007综合报告［M］. 瑞典：政府间气候变化专门委员会出版，2008.

［2］庄贵阳，陈迎. 国际气候制度与中国［M］. 北京：世界知识出版社，2005.

［3］联合国. 联合国气候变化框架公约，柏林授权，1995.

［4］联合国. 联合国气候变化框架公约，京都议定书，1998.

［5］国家发展改革委应对气候变化司. 清洁发展机制读本［M］. 北京：中国标准出版社，2008.

［6］国务院. 国务院关于成立国家应对气候变化及节能减排工作领导小组的通知，2007.

［7］中国清洁发展机制基金管理中心. 气候变化融资［M］. 北京：经济科学出版社，2011.

［8］谢飞，孟祥明，胡烨. 清洁发展机制：撬动发展中国家低碳经济杠杆［N］. 中国财经报，2010年1月21日（第4版）.

［9］中国清洁发展机制的冲动与尴尬. http://news.sina.com.cn/c/sd/2009-12-23/140519322122.shtml.

［10］国家发展和改革委员会文件. 国家发展改革委关于开展低碳省区和低碳城市试点工作的通知. http://www.sdpc.gov.cn/zcfb/zcfbtz/2010tz/t20100810_365264.htm.

［11］中华人民共和国国民经济和社会发展第十二个五年规划纲要. http://news.xinhuanet.com/politics/2011-03/16/c_121193916_12.htm.

［12］孟祥明，冯超，谢飞. 全球CDM市场发展及其面临的挑战［J］. 经济研究参考，2009（17）.

[13] 谢飞，孟祥明，刘淼. 全球碳市场：冷热不均　寻求突破 [N]. 中国财经报，2010年6月24日（第4版）.

[14] Carbon Finance at World Bank. State and Trends of the Carbon Market 2011.

[15] 谢飞，许明珠，孟祥明. 欧盟推出最新气候变化战略 [N]. 中国财经报，2010年4月8日（第4版）.

[16] 孟祥明，李春毅，谢飞. 2012年后，碳市场和CDM不会消失 [N]. 中国能源报，2011年1月24日（第 6 版）.

CDM and China CDM Fund

Edited by China CDM Fund

Editorial Committee

Foreword 1

Striding Ahead: Innovating Climate Change Financing Mechanism

Climate change financing has always been one of the focal points in global climate change negotiations and a key measure to tackle the climate crisis. Various countries have been endeavoring to innovate climate financing mechanism. China's actions have drawn worldwide attention.

Clean Development Mechanism (CDM) under Kyoto Protocol is one important product of this global endeavor. Through years of practice, CDM has been proved an effective way to promote sustainable development in developing countries and facilitate global response to climate change, leading to win-win benefits for developing and developed countries.

The Chinese government highly prioritizes climate change issues, and has taken practical actions to cope with it. China has actively taken part in international cooperation including CDM, explored suitable climate financing mechanisms, and supported domestic climate activities in a comprehensive, fast and efficient manner. Establishing China CDM Fund is one of the actions, and deemed a contribution of the Chinese government to climate financing innovation in the world.

The Chinese government collects national revenue from CER transactions. China CDM Fund uses the revenue to support national climate change strategies, policy making and implementation in mitigation and adaptation, elevate CDM from project-level cooperation to national-level in China, multiply the role of CDM and fill the gaps in its design. Meanwhile, China CDM Fund uses its funds to guide and leverage social capital to support domestic climate actions and facilitate the implementation of national strategies.

Currently, global climate change negotiations have come to a critical stage. Durban Climate Change Conference agreed to continue on a second commitment of Kyoto Protocol, but some core issues are still pending. Before the end of the first commitment period of Kyoto Protocol (2008–2012), it is imperative to solve these issues and make a seamless link between the two periods. Financing and technical issues will remain top priorities in follow-up negotiations.

The widely watched *12th-Five Year Plan for National Economic and Social Development* regards tackling climate change as a key to economic restructuring, and set up quantifiable targets on energy conservation and emission reduction. The *12th-Five Year Plan on GHG Emission Control* stipulated concrete implementing measures to reach the targets. This marks a new step on China's low carbon road, and points the way for public finance support to climate cause. It also provides great opportunity for China CDM Fund to play its role.

The book *CDM and China CDM Fund* gives a full picture of CDM projects in China and the world, summarizes China's successful experiences on CDM projects, and introduces in detail the establishment, governance, strategies, businesses, and operations of China CDM Fund. The book is a reference for global negotiations on climate change financing and also for China to further innovate climate financing; and also sets a good example on innovation of climate change financing mechanism to other countries, especially developing countries.

I expect China CDM Fund to make more contributions to global climate change.

Zhu Guangyao

Foreword 2

Enhancing the Sustainability of Innovative Climate Financing for Clean Energy

With its vision of Asia and the Pacific becoming free of poverty, the Asian Development Bank (ADB) has been actively supporting its developing member countries (DMCs) to pursue environmentally sustainable growth. Over the past two decades, remarkable achievements in economic growth have been made in the region. However, continued efforts for poverty reduction and improved quality of life in the region will not be sustainable without proactive efforts to mitigate and adapt to the negative impacts of climate change. As the only multilateral development bank headquartered in the region, ADB has been playing a key role in mobilizing innovative financing to assist clean energy projects and mitigation and adaptation actions.

The Clean Development Mechanism (CDM) is a key mechanism introduced by the Kyoto Protocol, to incentivize clean energy projects with certified emission reductions (CERs) in developing countries. CDM helps developing countries gain technology, funds and most importantly promotes low carbon development concepts. CDM is generally perceived as a win-win arrangement on carbon reduction between developed and developing countries. By 2011, about 3,900 projects, distributed over more than 70 countries, have been officially registered as the CDM projects triggering investments of more than USD 40 billion annually.

The People's Republic of China (the PRC) has been a very active participant in the CDM with the largest portfolio of registered projects. Since 2005, about 1,800 projects in the PRC have been registered. The successful uptake of CDM has helped the PRC utilize international resources effectively to implement its climate change-related programs. It not only helped the PRC to deploy cleaner technologies at an early stage to help combat climate change, it also mainstreamed the low-carbon development approach in a rapidly developing country. The initial momentum built by the CDM projects has helped the PRC become the leading country in the clean energy investment and manufacturing.

The sustainability of climate change financing has long been at the center of climate change policy discussion. The establishment of the China CDM Fund was one of the initiatives that the PRC government took to efficiently utilize its CDM resources. Through a levy on CER transactions, the China CDM Fund collect funds to reinforce national climate change research programs and leverage larger public and private financing to scale up climate change mitigation efforts. It is innovative financing platform, that enhances the sustainability of climate financing for clean energy project.

ADB has long been working closely with the PRC government in various mitigation and adaptation programs. Upon the request of the PRC government, ADB has provided since 2006 two technical assistance projects on a grant basis to support the establishment and enhance capacity of the China CDM Fund. I am very pleased to see that the China CDM Fund has made many encouraging achievements. I echo ADB President Mr. Haruhiko Kuroda who said at the inauguration on 9 November 2007, "China CDM Fund was a pioneering move-it uses income from international carbon market to eliminate domestic barriers and develop low-carbon economy".

ADB is pleased to fund the printing of the book titled "CDM and China CDM Fund". The book briefly documents the development course of the Kyoto Protocol and CDM, details the establishment and goals of China CDM Fund, and introduces the good experiences in developing and implementing CDM projects in PRC. With the strengthening of global efforts on climate change, the need for knowledge innovations and best-practice diffusion has assumed even greater importance. It is no doubt that the book can help disseminate the successful lessons learned through CDM development in the PRC to other developing countries, and help deepen the international community's understanding on global cooperation mechanism on climate change.

Currently, global climate change efforts are far behind the need for addressing climate change effectively. We can look back at the stimulating effect of the CDM projects in the PRC in rapidly expanding its renewable and clean energy portfolio and establishing leadership in clean energy arena. We certainly hope that the PRC's CDM success story can be replicated widely across the region.

ADB will continue its close cooperation with the PRC government to promote its climate change mitigation programs. I wish China CDM Fund a bright future and an active role in global climate change cooperation.

Stephen P. Groff

Vice President

Operations 2

Asian Development Bank

Preface

Clean Development Mechanism (CDM), is one of the three flexible compliance mechanisms under UNFCCC and Kyoto Protocol. It is the carbon trade activity most broadly participated by developing countries, including China.

China's CDM projects witness fast growth since 2005, and the number of projects and trade value account for over 50% of the world total. Meanwhile, China CDM Fund, established based on carbon trade under CDM, has also quickly expanded, and provided timely and strong support to China's action on climate change and low carbon development. This unprecedented financing innovation has received wide attention and applaud from the international community.

As China CDM Fund Management Center implements the Technical Assistance projects supported by Asian Development Bank (ADB), the project officers from ADB noted that, China has made remarkable achievements and gained rich experiences in CDM project development and implementation, and set an excellent example in establishment and operation of China CDM Fund, playing positive role in supporting China tackle climate change and promote low carbon development. It is significant to document these experiences and outputs, and share them with other developing countries.

We are grateful to ADB for supporting the publication of the book. Our gratitude goes to Yang Hongliang, senior climate change specialist of ADB, Sun Yuqing, Wang Ning, Wen Gang, Tu Yi, Li Lei and other colleagues from China CDM Fund Management Center for their valuable suggestions. And we are also indebted to Tian Chen from China CDM Fund Management Center, Zheng Lili from State Asset Management Co., ltd. of Chinese Academy of Social Sciences, Huo Yan from Xinhua News Agency, and Liu Zhen from Lenovo China (Beijing) for the translation and text-proofing of the English version.

Gratitude also goes to the Economic Science Press for their lots of meticulous work for the compilation and publication of the book.

Though contributors of the book have endeavored to showcase related outputs and experiences in the most thorough and accurate manner, there should still be some faults and omissions due to our limited vision and knowledge. All readers' corrections and suggestions are greatly encouraged and welcomed.

Editor

June, 2012

Contents

Part Ⅰ Clean Development Mechanism

Part II China Clean Development Mechanism Fund

Part Ⅰ

Clean Development Mechanism

Chapter I

Climate Change and UNFCCC

The greenhouse gas emissions generated by human activities has led to climate change across the world, exerting increasing impact upon human beings, society and ecological environment. To take effective measures to mitigate and adapt to climate change has become one of the most serious challenges for the whole world.

1.1 Climate Change

1.1.1 Causes of Climate Change

The earth's surface receives shortwave radiation from the sun, and sends longwave radiation to the space. The shortwave radiation could directly reach the surface through atmosphere, while the longwave radiation is absorbed by the greenhouse gases surrounding the earth. The process creates the greenhouse effect, and raises the temperature of the earth surface and lower atmosphere at an equilibrium temperature suitable for human being and other creature's growth. In the past millennia, the concentration of global greenhouse gases has been stable or increasing slowly. The balanced greenhouse effect makes the temperature of the earth surface close to be constant.

However, since 1750, particularly after the industrial revolution, the greenhouse gas emissions have increased dramatically due to accelerating industrialization of developed countries. According to the Intergovernmental Panel on Climate Change

(IPCC) Fourth Assessment Report (AR4), since industrial revolution, particularly from 1970 to 2004, the greenhouse gases from human activities have risen by 70%. CO_2, one of the most important greenhouse gases, has increased by 80%. In 2005, the atmospheric concentration of CO_2 (379ppm[1]) and CH_4 (1774ppb[2]) exceeded by far the natural range over the last 650,000 years[3].

The IPCC forecasted that if the fossil fuel still dominated the global energy structure even after 2030, the greenhouse gas emissions would grow by 25–90 percent from 2000 to 2030, and the global temperature would increase by 0.2℃ per ten years in the next 20 years. Even the concentration of the greenhouse gases and the aerosol remained the same as 2000 level, the world would get warmer with temperature increasing by 0.1℃ per year[4]. The situation of climate change was rather challenging.

1.1.2 Damages of Climate Change

If the greenhouse gas emissions continue to rise at current or a higher rate, the world will further warm up, and disasters will also increase. The impact includes[5]:

Rising Sea level: According to the assessment of IPCC, temperatures in excess of 1.9 to 4.6℃ warmer than pre-industrial sustained for millennia will lead to eventual melt of the Greenland ice sheet. This would raise sea level by 7 metres — comparable to 125,000 years ago.

Increasing natural disasters: In the northern hemisphere, winter will be shorter, colder and more humid, while summer will be longer, hotter and drier. The subtropical area will be drier, and the tropical area will be more humid; the snow covered area will reduce, with large permafrost regions melting. The ice-covered sea will shrink; Extreme climate, such as hot wave, heavy rainfall, etc, will take place more frequently; Rising temperature will lead to rapid evaporation, reduced rainfall and changing rainfall patterns in different regions; Regions of high latitude will face more rain, while the subtropical land area will experience less rain.

1 ppm: parts per million.

2 ppb: parts per billion.

3 IPCC. *Climate Change Report 2007* [M]. Sweden: IPCC, 2008: 2–5.

4 IPCC. *Climate Change Report 2007* [M]. Sweden: IPCC, 2008: 7.

5 IPCC. *Climate Change Report 2007* [M]. Sweden: IPCC, 2008: 8–13.

Impact on certain systems, industries and regions: the impact involves the ecological system of land, sea and coastal region, water resources of middle-latitude arid regions, arid tropical regions and regions relying on melting snow, agriculture of low latitude regions, coastal system of low-lying area, the health of people with low adaptability, the Arctic Pole, Africa, small islands and Asia.

IPCC pointed out that the climate change might bring some abrupt or irreversible changes. If global average temperature increase exceeded 1.5–2.5℃, 20–30 percent species might face bigger risk of extinction. If it was over 3.5℃, 40–70 percent species might experience extinction.

Based on the above-mentioned facts, the negative influence of climate change is quite challenging.

1.2 UNFCCC

1.2.1 The Establishment of UNFCCC

Though scientist Svante Arrhenius from Sweden put forward the concept of greenhouse gas (hereinafter referred as GHG) effect in the late 19th century, and many scientists gradually realized the problems brought by GHGs, still scientists seldom conducted systematic research on climate change. It was the human environment conference in Stockhom in 1972 that enabled people to study climate change. In late 1970s, scientists began to regard climate change as a potential serious problem. In 1988, World Meteorological Organization (WMO) and United Nations Environmental Programme (UNEP) jointly established United Nations Framework Convention on Climate Change (UNFCCC). The Toronto meeting held in the same year marked the beginning of the high-level debate on climate change. In December 1988, the United Nations (UN) assembly adopted a resolution to protect climate for this generation and the offspring. At the Second World Climate Conference of ministerial level in June 1990, representatives from European Union (EU) proposed for the first time the initiative to protect atmosphere and control CO_2, and finally included the climate change convention to the Ministerial Declaration in September meeting. At the United Nations Conferences on Environment and Development (UNCED) in Rio de Janeiro in June 1992, governments signed the UNFCCC, ushering in the international response to climate change[1].

1.2.2 Major Content of the UNFCCC

UNFCCC was the first global framework for international response to climate change.

UNFCCC stipulated the goal of stabilizing the GHG concentration to prevent the dangerous human interference with climate system. Meanwhile, the convention considered the fact that "developed countries dominated GHG emissions in the past and now; the developing countries' emission per capita was relatively low; the share of developing countries as a part of the global emissions would increase to satisfy social and economic demands". So it is proposed that all the countries should shoulder common but differentiated responsibilities, and carry out extensive cooperation and participate in effective and appropriate international actions in accordance with each country's ability and socioeconomic conditions.

UNFCCC also ruled that developed countries should take the lead in reducing emissions to the 1990 level, and provide capital and technologies to help developing countries tackle climate change. Such capital and technical assistance should be different from Official Development Assistance (ODA) and commercial technology transfer. The responsibility of developing countries was to draft national information inventory with list of GHG sources and sink as major content, and formulate national plan to mitigate and adapt to climate change. The extent to which developing countries comply with responsibilities would depend on the capital and technology transferred by developed countries[2].

---

1 Zhuang Guiyang, Chen Ying. *China and the Global Climate System*[M]. Beijing: World Affairs Press, 2005: 33–35.

2 Zhuang Guiyang, Chen Ying. *China and the Global Climate System*[M]. Beijing: World Affairs Press, 2005: 41–42.

UNFCCC was the most influential and meaningful legal document in the field of international environment and development, covering production, consumption, living styles, etc. However, since the UNFCCC is a general treaty that did not elaborate on GHG emissions reduction, it was difficult to put the convention into practice.

Chapter II

Kyoto Protocol and Clean Development Mechanism

To enhance the operability of UNFCCC and ensure the comprehensive implementation of the convention, the Conference of Parties to UNFCCC initiated new moves.

2.1 Kyoto Protocol

In 1995, the first Conference of Parties (COP) adopted the Berlin Mandate taking the commitments of developed countries and other parties as incomplete. The parties agreed to initiate a new process to promote actions for post 2000. One of the actions was to introduce a new protocol or another legal document to reaffirm the commitments of Annex I parties, but no new commitments for developing countries[1].

After 3 years of negotiation, COP 3 was held in Kyoto from 1st to 11th December 1997 with the purpose of setting a binding emissions reduction goal and deadline to strengthen efforts of developed countries to reduce emissions and curb global warming. The emission reduction of developed countries was the central issue of the meeting. Kyoto Protocol was passed after difficult negotiation. It was seen as a milestone of addressing climate change globally, and a critical step for the implementation of UNFCCC. The meeting marked international negotiation on

1 UNFCCC, Berlin Mandate, 1995.

climate change entered a constructive stage. Meanwhile, Kyoto Protocol became the first legal document with specific obligation of quantified reduction, and laid the foundation for implementing UNFCCC.

2.1.1 Major Content of Kyoto Protocol

The Kyoto Protocol set 6 greenhouse gases (namely CO_2, Methane(CH_4), Nitrous oxide(N_2O), Hydrofluorocarbons (HFCs), Perfluorocarbons(PFCs) and Sulphur hexafluoride(SF_6)) as control targets according to related principles of UNFCCC and Berlin Mandate[1]. Meanwhile, the protocol also stipulated that Annex 1 parties should at least reduce emission over the first commitment period (from 2008 to 2012) by 5% below 1990 level.

Meanwhile, to reduce the emission reduction cost of developing countries and realize the emission reduction goal, Kyoto Protocol set up 3 flexible mechanisms: Joint Implementation (JI), Clean Development Mechanism (CDM) and Emissions Trading (ET).

Kyoto Protocol "shall enter into force on the ninetieth day after the date on which not less than 55 Parties to the Convention, incorporating Parties included in Annex I which accounted in total for at least 55 per cent of the total carbon dioxide emissions for 1990 of the Parties included in Annex I, have deposited their instruments of ratification, acceptance, approval or accession"[2].

2.1.2 The Development Course of Kyoto Protocol

Though the Kyoto Protocol was adopted in 1997, still it only involved emission reduction goals and mechanisms without technical details, leading to several rounds of tough negotiation. In November 2004, Russia signed Kyoto Protocol, which marked that the emissions of signing parties took up over 55% of CO_2 emissions of the Annex I parties. On 16th February 2005, the Kyoto Protocol entered into force. By the end of 2010, 192 countries signed the protocol, representing 63.7% of the CO_2 emissions of

1 Global Warming Potential (GWP) of Six Greenhouse Gases: CO_2, 1; CH_4, 21; N_2O, 310; HFCs, 140–11,700; PFCs, 6,500–9,200; SF_6, 23,900.

2 UN. UNFCCC, Kyoto Protocol, 1998.

Annex Ⅰ parties for 1990[1].

The COP 4 was held in Buenos Aires, Argentina from 2nd to 13th November 1998. The conference adopted Buenos Aires Plan of Action that reiterated the principle of common but differentiated responsibility. The Plan ruled that at the sixth session, agreement should be reached concerning enforcement rules, and commitments of providing financial and technological support to developing countries should be promoted, paving the way for the effectiveness of Kyoto Protocol in 2002. The Plan guided the direction of Kyoto Protocol for deeper negotiations.

The COP 5 was held in Bonn, Germany from 25th October to 5th November 1999. At this meeting, participants discussed about specific rules concerning the effectiveness of protocol, including developing countries' participation, Kyoto mechanism, compliance procedures, carbon sink, etc.

The COP 6 was held in Hague, Netherlands from 13th to 24th November 2000. The purpose was to formulate specific measures of implementing Kyoto Protocol to enable developed countries to fulfill commitments of emission reduction. Because of the conflict of countries' interests, the meeting made tardy progress. What should be mentioned particularly were the divergence between EU and umbrella group led by US on carbon trade, carbon sink and performance mechanism, the opposite views of developed countries and developing countries on technology development and transfer, capacity building and capital mechanism, and the difference among developing countries on some issues. The meeting did not reach agreement on many issues, and the follow-up meeting was scheduled in 2001 for further negotiations.

The sixth session re-opened in Bonn, Germany from 16th to 27th July 2001, at which Bonn Agreement was adopted. Before the session, IPCC published the third assessment report in January 2001, and former US president George W Bush pulled US out of the Kyoto Protocol in March 2001. These events made people realize the significance and complexity of climate change, and the difficulty of relevant negotiations. Through tough negotiations and compromises, the follow-up session

1 UNFCCC. Status of Ratification of the Kyoto Protocol. http://unfccc.int/essential_background/kyoto_protocol/status_of_ratification/items/5524.php.

made initial agreement on the fund mechanism, technology development and transfer, Kyoto Protocol's participating qualifications, application scope, implementing organization, compliance mechanism, land use, land-use change and forestry (LULUCF), etc. The Bonn Agreement was the product of compromise. Developing countries made big concessions and developed countries alleviated their burden of reducing emissions and providing technical and economic support, laying the political foundation for the final effectiveness of Kyoto Protocol. All parties, without the participation of US, agreed upon the enforcement rules under the principle of common but differentiated responsibility, implying a breakthrough of international climate system.

The COP 7 was convened in Marrakech, Morocco from 29th October to 10th November 2001. The meeting adopted the decisions of Bonn Agreement as a package called Marrakech Accords. Specifically, the meeting passed the draft decisions of Bonn Agreement on capital, technical transfer, capital building, etc. and reached a package deal concerning the three mechanisms, carbon sink and compliance procedures shelved by follow-up session of COP 6. Moreover, developed countries made progress in financial assisstance. The meeting brought Kyoto Protocol to a critical stage where parties were ready to approve the convention. Marrakech Accords implied the transfer from theory to practice, from international negotiation to global action.

The COP 8 was convened in New Delhi, India from 23rd October to 1st November 2002. The meeting revolved around climate change under the framework of sustainable development, and proposed that adaptation measures should be the priority of all countries. Delhi Declaration was passed at this session. According to the principle of common but differentiated responsibility, to consider climate change under the framework of sustainable development became a key to all countries' climate change strategies.

The COP 9 was held in Milan, Italy from 1st to 12th December 2003. At the meeting, EU endeavored to promote global commitments of emission reduction, and encourage developing countries to shoulder substantive responsibilities. Russia expressed its disapproval of Kyoto Protocol for the time, further postponing the effectiveness of Kyoto Protocol.

The COP 10 was held in Buenos Alice, Argentina from 6th to 17th December 2004. The year marked the 10th anniversary of UNFCCC. Russia signed Kyoto Protocol a month before the meeting, making the session the last one before the final effectiveness of Kyoto Protocol. A series of adaptation measures were adopted at the meeting. Considering the impact of climate change, the parties approved the Buenos Aires Adaptation Work Plan which included many seminars and technical documents concerning climate change risks and adaptation. The plan held that adaptation should be a key part of the sustainable development, and supported the "National Adaptation Programmes of Action (NAPAs)" aimed at the least developed countries. The meeting called for countries to take measures to fulfill commitments to UNFCCC and Kyoto Protocol[1].

2.2 Clean Development Mechanism

2.2.1 Basic Concept of CDM

CDM is one of the three flexible mechanisms under Kyoto Protocol. It is intended to meet two objectives: (1) to assist parties not included in Annex I in achieving sustainable development and in contributing to the ultimate objective of UNFCCC; and (2) to assist parties included in Annex I in achieving compliance with their quantified emission limitation and reduction commitments[2].

The core of CDM was that countries included in Annex I parties should carry out emission reduction program with developing countries by providing fund and technology. The Certified Emission Reductions (CERs) could be used by developed countries to meet their commitments.

Kyoto Protocol stipulated that CDM could be implemented after 2000, and the accumulated CERs could allow Annex I countries to meet part of their responsibilities

1 Climate Change Department of NDRC. *Clean Development Mechanism* [M]. Beijing: Standards Press of China, 2008: 4-6.

2 UN. UNFCCC, Kyoto Protocol, 1998.

in the first commitment period from 2008 to 2012. Developing countries had the right to determine the priority area of CDM projects according to their sustainable development strategy. Parties to the CDM project could proceed on a voluntary basis with the approval of bilateral governments as the prerequisite.

CDM was generally perceived as a win-win mechanism. Theoretically, developed countries could meet their emission reduction objective in a cost-effective way, and at the same time bring in technology, products and clean development concept to developing countries. Developing countries could gain technology, concept, funds and even more investment to promote economic development, environmental protection and achieve the goal of sustainable development.

Many technologies and measures were applicable to CDM projects. In a broad sense, any technology and measures suitable for emission reduction and GHG recycling or capture could apply for a registered CDM project. According to the United Nations CDM Executive Board (EB), CDM projects mainly focus on the following 15 industries[1].

(1) Energy industry (renewable-/non-renewable energy) ;

(2) Energy distribution;

(3) Energy demand;

(4) Manufacturing industries;

(5) Chemical industries;

(6) Building;

(7) Transport;

(8) Mining/ mineral production;

(9) Metallurgy;

(10) Fugitive emissions from fuels(solid, oil and gas);

(11) Fugitive emissions from production and consumption of halocarbons and sulphur hexafluoride;

(12) Solvent use;

(13) Waste handling and disposal;

(14) Afforestation and reforestation;

---

1 CDM Executive Board of UN. http://cdm.unfccc.int/Statistics/Registration/RegisteredProjByScopePieChart.html.

(15) Agriculture.

2.2.2 CDM's International Rules and Requirements

CDM was only defined in principle in Article 12 of Kyoto Protocol. Since 1998, the COP, the Subsidiary Body for Science and Technological Advice (SBSTA), the Subsidiary Body for Implementation (SBI) had been formulating rules for CDM cooperation. After 4 years of tough negotiation, agreement was finally reached on enforcement rules at COP 7 in 2001. These rules were mainly reflected in the 15th and 17th resolution and relevant attachments. CDM EB made a series of decisions on many detailed technical problems[1].

2.2.2.1 Qualifications for Participation

The 17th document of COP 7— *Modalities and Procedures of CDM* specified on the qualifications of participating in CDM projects in paragraphs 28 to 32. Only the contracting parties can take part in CDM project cooperation. Moreover, participation must be on voluntary basis and must involve the Designated National Authorities (DNA) responsible for CDM issuance.

Chinese State Council ratified Kyoto Protocol in 2002. After it entered into force, China became one of the participants of CDM on a legal basis.

2.2.2.2 Project Eligibility Criteria

Paragraphs 37 to 52 of *Modalities and procedures of CDM* stipulated many requirements of eligibility. The requirements mainly included: projects must reduce GHG emissions compared to baseline; projects must be approved by parties' DNA; The methodologies must be approved; if projects cause other environmental issues, there must be corresponding countermeasures; the baseline must be project-based and built in a conservative and transparent way; it should also consider national and industrial policies and rules; rational boundary should be set for projects; GHG leakage should also be taken into account.

---

1 Climate Change Department of NDRC. *Clean Development Mechanism* [M]. Beijing: Standards Press of China, 2008: 6–8.

China mainly focused on the following three factors:

(1) Project baseline and CER. Projects must reduce GHG emissions.

(2) Projects must bring technology transfer. It can be foreign technologies or domestic ones with low degree of commercialization or difficult to be commercialized.

(3) Funds from projects. The funds should be provided by developed countries, and should be separated from ODA.

2.2.2.3 Major Parties

Major participants of CDM projects include project owner, host government, designated operational entity (DOE) for certification/verification, CDM EB and COP.

(1) Project owner. Project owner works out project design document (PDD) in accordance with the format issued by EB, and submits it to the host government for approval while inviting a DOE to validate the project. After validation and registration, the project owner will implement and monitor the project according to PDD. After a period, another DOE will be invited to verify the GHG emission reduction.

(2) Host government. It is in charge of examining whether CDM project conforms to national sustainable development needs and related policies, and deciding whether to approve the CDM project. Host government could manage CDM projects developed by domestic and foreign institutions through issuing policies and establishing specialized agencies.

(3) DOE. The major responsibility is to validate project according to CDM rules, and submit to EB for approval and registration; after operating the project, DOE verifies GHG emission reductions, and applies to EB for the issuance of CERs.

(4) CDM EB. The main responsibility is to regulate the implementation of CDM project. It mainly includes: formulating the implementation rules for CDM project according to the decisions and guidelines of COP; putting forward and approve CDM project methodologies; appointing the DOE and submitting to the COP for approval; examining and approving the application for CDM project registration; issuing the CERs of CDM project.

(5) COP. COP, which is the top decision-making body of CDM, decides all

the issues of UNFCCC and Kyoto Protocol. The conference, joined by all parties, is convened once a year and discusses all issues on climate change.

2.2.2.4 Project Development Process

A typical CDM project development process is shown in diagram 2-1.[1]

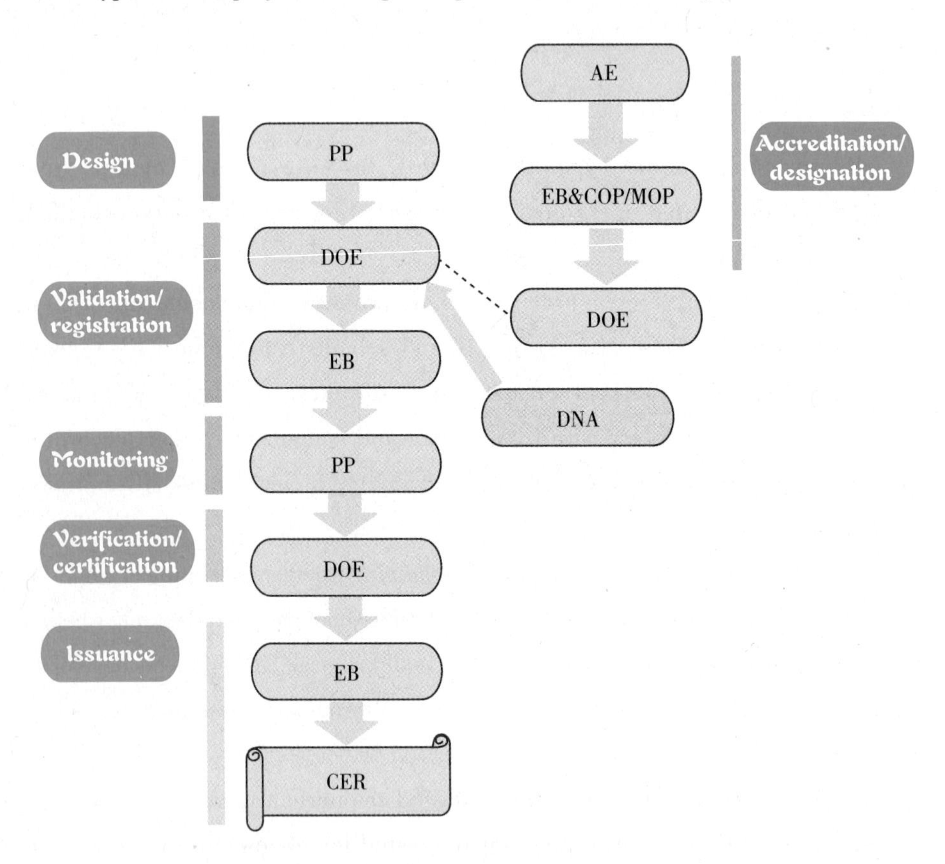

Diagram 2-1 A typical CDM project development process

(1) Project preparation. Project owner formulates PDD according to the format provided by EB, and submits it to the host government for approval. Considering the technicality of PDD and English as the working language, currently in China, the project owner usually entrusts professional consultancy with the task of preparing

1 Climate Change Department of NDRC. *Clean Development Mechanism* [M]. Beijing: Standards Press of China, 2008, 32-34.

PDD. Some capable project owners formulate the paper on their own.

(2) Government approval. DNA organized experts to examine and issue the Letter of Approval (LOA) to those qualified CDM projects.

(3) Project validation. After government's approval, project owner chooses a EB-authorized DOE to validate the project. The two sign a contract and the project owner provides relevant materials to the DOE to validate the project according to CDM rules; after validation, DOE will submit the validation report to EB for registration.

(4) Project registration. The secretariat of UNFCCC examines the integrity of PDD, and publishes the project on the official website of UNFCCC under the column of CDM for 8 weeks after ensuring the integrity. During the period, if at least 3 members of EB or any project participant requested for review, the project could not be registered.

(5) Project monitoring. After registration, project owner starts operation, and monitors the project implementation according to the PDD, particularly the GHG emission reductions.

(6) Verification/certification of emission reduction. After a period, project owner invites another DOE to verify the emission reduction. DOE, in accordance with monitoring plan, formulates the project verification and certification report including the actual emission reductions, and submits it to EB for CER issuance.

(7) CER issuance. The secretariat of UNFCCC examines the integrity of CER application, and publishes the application on the website of UNFCCC under the column of CDM for 15 days. During the period, if 3 or more members of EB or any other project participant requested for review, the CERs will not be issued.

(8) CER transfer. After CERs issuance , the CERs would be put into the account of host country temporarily. After confirmation of buyer and seller, 2% of the CER would be donated to UN climate change adaptation fund, and the rest would be transferred to the buyer's designated account.

2.3 China CDM Project Management

2.3.1 China's Actions on Climate Change and the Management Framework

As a responsible developing country, China has payed high attention to climate change. On one hand, China played a positive and constructive role in promoting international negotiations under the principle of UNFCCC and Kyoto Protocol. On the other hand, China has honored its commitments to the international society in an all-round way.

China signed the UNFCCC at Unite Nations Conference on Environment and Development in Rio de Janeiro on 11th June 1992, and officially ratified the convention on 17th November 1992, making it one of the first parties. China signed and ratified Kyoto Protocol on 29th May 1998 and 30th August 2002 respectively, becoming the 37th signatory[1]. Since then, China has been implementing its duties comprehensively.

On 12th June 2007, the State Council established the National Leading Group on Climate Change including premier Wen Jiabao and over 20 ministers. As a national discussion and coordination institution, the group formulated important strategies and principles on climate change, planned work on climate change, discussed countermeasures in international negotiations, coordinated on important issues of climate change, implemented policies on energy conservation and emission reduction, etc[2].

Meanwhile, China formulated *China's National Climate Change Program* and *China's Policies and Actions for addressing Climate Change*. China also set

1 UNFCCC. Status of Ratification of the Kyoto Protocol. http://unfccc.int/essential_background/kyoto_protocol/status_of_ratification/items/5524.php.

2 State Council. Circular on establishing National leading group on climate change and energy conservation and emission reduction. June, 12th 2007.

goals of energy conservation and emission reduction, and integrated tackling climate change into the national development plan.

2.3.2 China CDM Project Management System

China attached great importance to CDM since the very beginning, which helped us utilize international resources effectively to address climate change.

On 30th June 2004, China issued *the Interim Measures for Management of CDM Project Operation*. After one year's trial implementation, China promulgated the *Measures for Management of CDM Project Operation* in China which stipulated the permissive conditions of CDM project application, national examination and approval, management and implementation procedures. This ensured the standardization and consistency of project operation. On 3rd August 2011, National Development and Reform Commission (NDRC), Ministry of Science and Technology (MOST), Ministry of Foreign Affairs (MOFA) and Ministry of Finance (MOF) jointly revised the *Measures* and issued the revised edition to promote the healthy development of CDM projects.

Meanwhile, in order to strengthen management of CDM project to ensure project quality and safeguard national interests and reputation, China established the National CDM Board under the National Coordination Committee on Climate Change. NDRC and MOST served as co-chairs of, and MOFA served as vice chair of the Board, which also includes Ministry of Finance (MOF), Ministry of Environmental Protection (MEP), Ministry of Agriculture(MOA). China Meteorological Administration (CMA), and The Board's responsibilities include: reviewing and commenting on CDM projects; reporting to the Committee implementation of CDM projects, possible problems and proposals; giving advices on national CDM project rules. As the Designated National Authority of CDM projects, the responsibilities of NDRC include: accepting project application; approving CDM project activities jointly with MOST and MOFA according to the review result of the Board; issuing Letter of Approval; organizing supervision on the implementation of project activities. The national CDM management framework was shown in diagram 2–2.

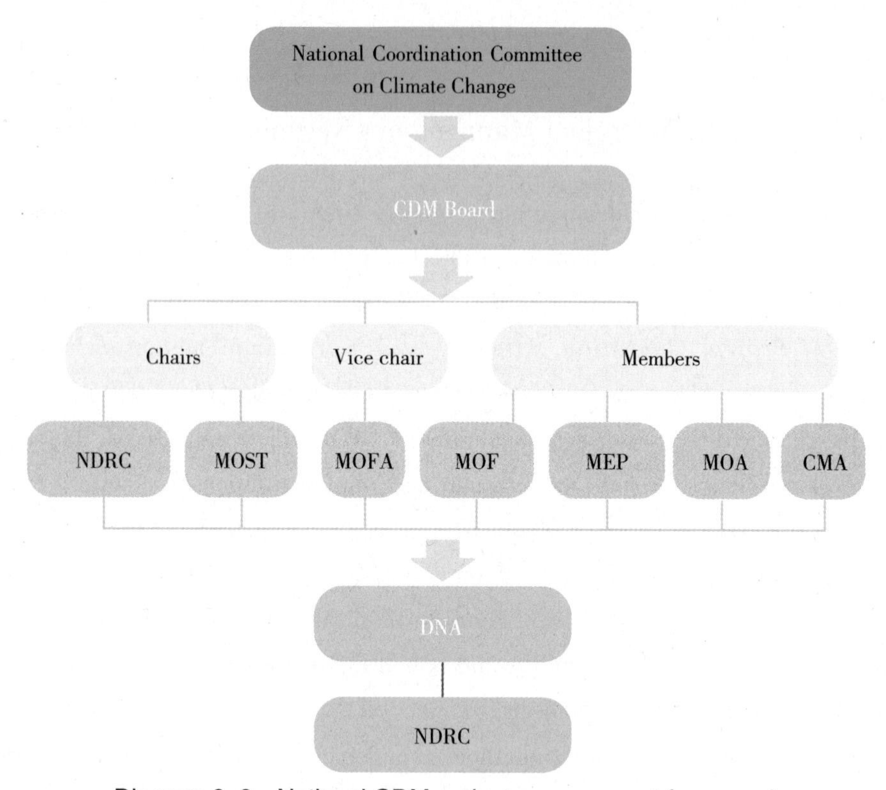

Diagram 2–2 National CDM project management framework

Chapter Ⅲ

Current Status of CDM

Among the three flexible mechanisms, CDM was the only one that linked developed countries and developing countries. Since 2004, with several years' development, CDM has become the most mature mechanism of jointly addressing climate change.

3.1 Status of Global CDM Project Development

According to relevant CDM rules, the CDM project development and implementation has to go through national approval, registration at the CDM EB and issuance of CERs by EB before it finally come to CERs trading. We could understand the general situation of CDM projects through analyzing the three procedures. Since the national approval is a domestic procedure and it was difficult to obtain the detailed statistics of every country through public channels, so analysis on global CDM project development focuses more on project registration and CERs issuance.

3.1.1 Project Registration

3.1.1.1 The Status of Project Registration

By 30th June 2011, 3,368 CDM projects, distributed in 71 countries, have been registered by EB. The expected annual emission reductions will reach 507 million ton CO_2 equivalent (tCO_2e) if all the projects implemented smoothly.

According to the statistics of project number, these projects were mainly distributed in China, India, Brazil, Mexico, Malaysia and Indonesia. The projects in the above 6 countries accounted for 81.2% of the global registration number. Among the 6 countries, China led the way with its share of 45.0%. The distribution of global registered projects was shown in table 3–1.

Table 3–1 Distribution of global registered project number

Country	Number of Projects	Percentage (%)
China	1,516	45.0
India	709	21.0
Brazil	212	6.3
Mexico	130	3.9
Malaysia	97	2.9
Indonesia	70	2.1
Others	634	18.8
Aggregate	3,368	100.0

Source: Data collected from the website of CDM EB, UNFCCC.

Measured by expected annual emission reductions, these projects were mainly located in China, India, Brazil, South Korea, Mexico and Indonesia. The expected annual emission reductions of the 6 countries accounted for 86.3% of the total reductions. China led the way with its share of 63.5%. The distribution of the expected annual emission reductions of global registered projects was shown in table 3–2.

Table 3–2 Distribution of global projects' expected annual emission reductions

Country	Expected Annual Emission Reductions ($MtCO_2e$)	Percentage (%)
China	318.67	63.5
India	53.02	10.6
Brazil	24.31	4.8

Renewal table

Country	Expected Annual Emission Reductions ($MtCO_2e$)	Percentage (%)
South Korea	18.72	3.7
Mexico	10.49	2.1
Indonesia	7.53	1.5
Others	68.70	13.7
Aggregate	507.23	100.0

Note: $1MtCO_2e$=1,000,000 tCO_2e.

Source: Data collected from the website of CDM EB, UNFCCC.

3.1.1.2 The Trend of Registered Projects

Since the first CDM project was registered on 18th November 2004, the number of registered projects has been increasing quickly. After 1,195 days, on 26th February 2008, the number exceeded 1,000. On 13th December 2009, the number surpassed 2,000 within 656 days. On 2nd February 2011, the number topped 3,000 within 416 days. By the end of 30th June 2011, the project number reached 3,369. If increasing at such a speed, the project number will take about 400 days to reach 4,000. The time taken to exceed 1,000 projects is shown in diagram 3-1.

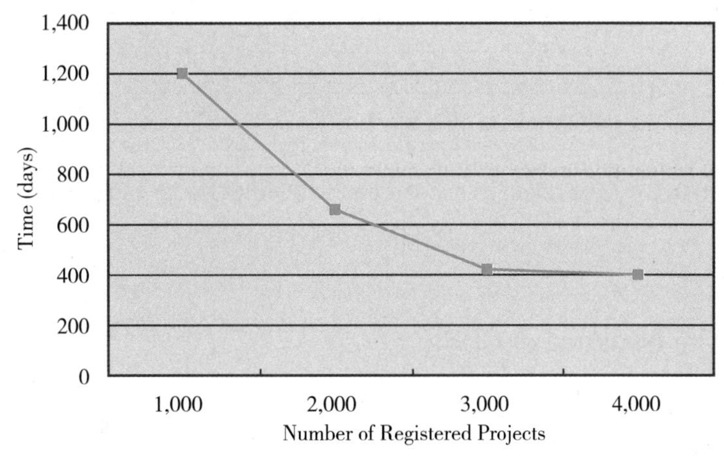

Diagram3-1 Time taken for registered projects to exceed 1,000

Source: Data collected from the website of CDM EB, UNFCCC.

On 27th October 2006, 708 days after the first project registered on 18th November 2004, the expected annual emission reductions of global registered projects exceeded 100 $MtCO_2e$ for the first time. On 1st February 2008, the reduction topped 200 $MtCO_2e$ within 462 days. On 20th April 2009, the reduction overtook 300 $MtCO_2e$ within 444 days. On 16th October 2010, the reduction exceeded 400 $MtCO_2e$ within 544 days. On 28th April 2011, the reduction was over 500 $MtCO_2e$ within 194 days. The time taken to exceed 400 $MtCO_2e$ was longer than before mainly because: most industrial gas projects like HFC-23 with higher annual emission reductions have come to the finish point, leaving us with relatively small scale projects in other areas; plus CDM EB of UN and DOEs had efficiency problems. The time taken to exceed 500 $MtCO_2e$ of annual emission reductions for registered projects was much shorter because of the greatly improved efficiency of EB. Based on current situation, the time taken to exceed 600 $MtCO_2e$ will continue to shrink. The time taken to exceed each 100 $MtCO_2e$ was shown in diagram 3–2.

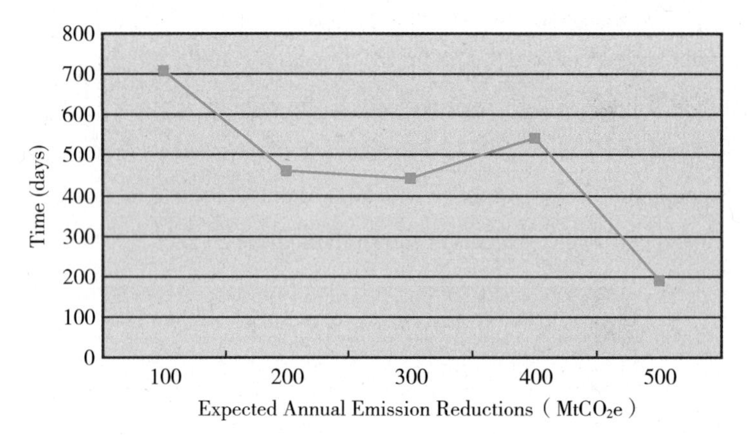

Diagram 3–2 Time taken to exceed each 100 $MtCO_2e$ for registered projects

Source: Data collected from the website of CDM EB, UNFCCC.

3.1.2 The Issuance of CERs

3.1.2.1 The Status of CERs Issuance

By 30th June 2011, all around the world, 2,789 batches of CERs which were equivalent to 647 $MtCO_2e$, have been issued. Estimated that the CER's transaction price was averaged as 8 USD/tCO_2e, CDM has delivered USD 5.2 billion to developing countries for sustainable development[1]. These CERs were mainly generated in 44 developing countries, among which China, India, South Korea, Brazil, Mexico and Chile account for 93.9% of total issuance. China ranked first in the whole world with its 368 $MtCO_2e$ or 56.9% of global issuances. The distribution of global CER issuances was shown in table 3–3.

Table 3–3 Distribution of global CERs

Country	Issuance Batch		Issuance Amount	
	Batch	Proportion（%）	($MtCO_2e$)	Percentage（%）
China	1,125	40.3	368.07	56.9
India	672	24.1	97.76	15.1
South Korea	105	3.8	71.79	11.1
Brazil	389	13.9	54.10	8.4
Mexico	125	4.5	8.34	1.3
Chile	55	2.0	7.08	1.1
Others	318	11.4	39.59	6.1
Aggregate	2,789	100.0	646.73	100.0

Note: 1 $MtCO_2e$=1,000,000 tCO_2e.

Source: Data collected from the website of CDM EB, UNFCCC.

3.1.2.2 The Trend of CERs Issuance

The first CERs was issued on 20th October 2005.On 14th December 2007, the issued CERs exceeded 100 $MtCO_2e$ within 785 days. On 16th October 2008, the issued CERs topped 200 $MtCO_2e$ within 307 days. On 23rd June 2009, the issued CERs were over 300 $MtCO_2e$ within 250 days. On 9th April 2010, the issued CERs overtook 400 $MtCO_2e$ within 290 days. On 10th January 2011, the issued CERs were over 500 $MtCO_2e$ within 276 days. On 27th April 2011, the issued CERs were over 600 $MtCO_2e$ within 107 days. By 30th June 2011, the issuance reached 685 $MtCO_2e$

1 The earlier price for transaction was lower, being 6–8 dollars per tCO_2e; price grew higher later.

within 64 days. The time taken to exceed 700 $MtCO_2e$ would be shorter. The time taken to exceed each 100 $MtCO_2e$ was shown in diagram 3–3. It showed that the issuance time was speeding up as project proceed. Generally speaking, the time taken to exceed each 100 $MtCO_2e$ was decreasing, except the time taken to top 400 and 500 $MtCO_2e$. This was mainly due to the fact that EB suspended the issuance of all HFC-23 decomposition projects in the latter half of the year 2010, and resumed in December 2010, the related issuance was extended to the first half of the year 2011.

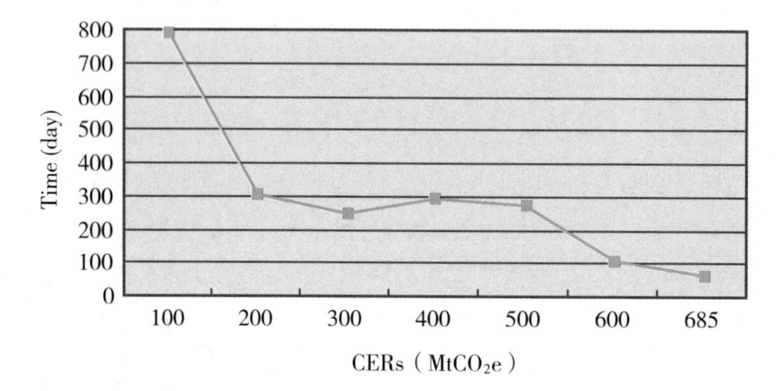

Diagram 3–3 Time taken to exceed each 100 $MtCO_2e$ for CERs issuance

Note: 1 $MtCO_2e$=1,000,000 tCO_2e.

Source: Data collected from the website of CDM EB, UNFCCC.

Distribution of global CDM projects was shown in table 3–4.

Table 3–4 Global CDM projects distribution

Country	Registration				CERs Issuance			
	Number of Projects	Percentage (%)	Expected Annual Emission Reductions(tCO_2e)	Percentage (%)	Batch	Percentage (%)	Issuance Amount (tCO_2e)	Percentage (%)
China	1,516	45.0	318,665,043	63.5	1,125	40.3	368,066,457	56.9
India	709	21.0	53,023,555	10.5	672	24.1	97,758,431	15.1
Brazil	212	6.3	24,312,766	4.8	389	13.9	54,101,186	8.4
Mexico	130	3.9	10,490,396	2.1	125	4.5	8,340,029	1.3
Malaysia	97	2.9	5,589,842	1.1	22	0.8	1,216,896	0.2
Indonesia	70	2.1	7,532,212	1.5	21	0.8	2,604,760	0.4
Vietnam	67	2.0	3,494,109	0.7	5	0.2	6,646,339	1.0

Renewal table

Country	Registration				CERs Issuance			
	Number of Projects	Percentage (%)	Expected Annual Emission Reductions(tCO_2e)	Percentage (%)	Batch	Percentage (%)	Issuance Amount (tCO_2e)	Percentage (%)
South Korea	60	1.8	18,723,184	3.7	105	3.8	71,790,390	11.1
Philippine	55	1.6	2,158,700	0.4	7	0.3	240,036	0.0
Thailand	54	1.6	3,139,308	0.6	5	0.2	851,541	0.1
Chile	50	1.5	5,707,230	1.1	55	2.0	7,077,779	1.1
Columbia	32	0.9	3,643,079	0.7	21	0.8	940,317	0.1
Peru	24	0.7	2,757,210	0.5	14	0.5	609,611	0.1
Argentina	23	0.7	4,917,604	1.0	36	1.3	6,478,615	1.0
Israel	22	0.7	2,223,049	0.4	17	0.6	866,907	0.1
Honduras	21	0.6	371,572	0.1	27	1.0	522,629	0.1
South Africa	19	0.6	3,247,426	0.6	13	0.5	1,900,276	0.3
Ecuador	16	0.5	1,371,456	0.3	22	0.8	1,114,540	0.2
Pakistan	12	0.4	1,774,587	0.3	11	0.4	2,982,626	0.5
Guatemala	11	0.3	864,760	0.2	15	0.5	947,952	0.1
Uzbekistan	11	0.3	4,402,064	0.9	0	0.0		0.0
Egypt	9	0.3	3,068,050	0.6	16	0.6	6,367,204	1.0
Costa Rica	8	0.2	418,606	0.1	6	0.2	320,463	0.0
Panama	7	0.2	358,513	0.1	1	0.0	60,180	0.0
Cyprus	7	0.2	300,889	0.1	0	0.0		0.0
Sri Lanka	7	0.2	210,168	0.0	8	0.3	237,690	0.0
Saldivar	6	0.2	619,535	0.1	8	0.3	790,253	0.1
Uruguay	6	0.2	354,713	0.1	1	0.0	40,613	0.0
Iran	6	0.2	692,684	0.1	0	0.0	0	0.0
The United Arab Emirates	5	0.1	356,416	0.1	1	0.0	79,960	0.0
Cambodia	5	0.1	150,948	0.0	1	0.0	10,758	0.0
Kenya	5	0.1	1,201,980	0.2	0	0.0		0.0
Morocco	5	0.1	287,447	0.1	4	0.1	330,099	0.1

Renewal table

Country	Registration				CERs Issuance			
	Number of Projects	Percentage (%)	Expected Annual Emission Reductions(tCO_2e)	Percentage (%)	Batch	Percentage (%)	Issuance Amount (tCO_2e)	Percentage (%)
Nicaragua	5	0.1	585,296	0.1	10	0.4	577,757	0.1
Nigeria	5	0.1	4,693,552	0.9	1	0.0	1,867	0.0
Armenia	5	0.1	223,063	0.0	1	0.0	12,022	0.0
Bolivia	4	0.1	563,991	0.1	3	0.1	1,117,802	0.2
Moldova	4	0.1	226,585	0.0	0	0.0	0	0.0
Nepal	4	0.1	154,317	0.0	0	0.0	0	0.0
Uganda	4	0.1	117,550	0.0	0	0.0	0	0.0
Syria Arab Republic	3	0.1	320,782	0.1	0	0.0	0	0.0
Ivory Coast	3	0.1	639,203	0.1	0	0.0	0	0.0
Rwanda	3	0.1	29,682	0.0	0	0.0	0	0.0
Mongolia	3	0.1	71,904	0.0	1	0.0	48	0.0
Paraguay	2	0.1	18,711	0.0	0	0.0	0	0.0
The Kingdom of Bhutan	2	0.1	499,522	0.1	1	0.0	474	0.0
Dominican Republic	2	0.1	483,726	0.1	1	0.0	11,637	0.0
Fiji	2	0.1	47,399	0.0	2	0.1	35,550	0.0
Congo	2	0.1	179,330	0.0	0	0.0		0.0
Georgia	2	0.1	411,897	0.1	1	0.0	53,138	0.0
Cuba	2	0.1	465,397	0.1	2	0.1	171,178	0.0
Cameroon	2	0.1	193,462	0.0	0	0.0	0	0.0
Bengal	2	0.1	169,259	0.0	0	0.0	0	0.0
Tunisia	2	0.1	687,573	0.1	0	0.0	0	0.0
Singapore	2	0.1	116,782	0.0	0	0.0	0	0.0
Jordan	2	0.1	434,074	0.1	4	0.1	985,992	0.2
Albania	1	0.0	22,964	0.0	0	0.0		0.0
Ethiopia	1	0.0	29,343	0.0	0	0.0		0.0
Papua New Guinea	1	0.0	278,904	0.1	2	0.1	215,424	0.0

Renewal table

Country	Registration				CERs Issuance			
	Number of Projects	Percentage (%)	Expected Annual Emission Reductions(tCO_2e)	Percentage (%)	Batch	Percentage (%)	Issuance Amount (tCO_2e)	Percentage (%)
Guyana	1	0.0	44,733	0.0	0	0.0	0	0.0
Qatar	1	0.0	2,499,649	0.5	0	0.0	0	0.0
Laos	1	0.0	3,338	0.0	1	0.0	2,168	0.0
Lybia	1	0.0	93,635	0.0	0	0.0	0	0.0
Madagascar	1	0.0	44,196	0.0	0	0.0	0	0.0
Mali	1	0.0	188,282	0.0	0	0.0	0	0.0
Republic of Macedonia	1	0.0	54,623	0.0	0	0.0	0	0.0
Senegal	1	0.0	37,386	0.0	0	0.0	0	0.0
The United Republic of Tanzania	1	0.0	202,271	0.0	2	0.1	35,122	0.0
Jamaica	1	0.0	52,540	0.0	4	0.1	211,223	0.0
Zambia	1	0.0	130,032	0.0	0	0.0		0.0
Aggregate	3,369	100.0	507,233,736	100.0	2,789	100.0	646,725,939	100.0

Note: The statistics were by the end of 30 June 2011.

Source: Data collected from the website of CDM EB, UNFCCC.

Development Speed of Global CDM Projects

Diagram 3-4 below showed the development of global CDM projects registration on a six months basis. From 2004 to 2006, global CDM projects developed very fast. In this period, the project number was relatively small, so the EB's work was not yet saturated. However, since the latter half of 2007, the registration slowed, with the number decreasing from 250 in the latter half of 2006 and first half of 2007 to 174, and stabilizing at 200~300 afterwards.

As countries emphasized more on CDM projects and proficiency of project development increased, the registration number grew rapidly, and the work efficiency of EB became a constraint. To improve work efficiency, EB began to opitimize procedures of examination and approval in the latter half of 2010. Moreover, in Cancun Agreement of the 2010 United Nations Climate Change Conference, EB issued "further guidance relating to clean development mechanism" which further define and optimize CDM project procedures, set timeline for project approval and improved efficiency of each step. Due to the optimization, registration increased remarkably to 548 in the latter half of 2010 and 517 in the first half of 2011, and the registered projects in 6 months jumped from 200-300 to over 500.

With the implementation of CDM project, CERs issuance tended to be saturated, so the speed of issuance became relatively stable. In 2010, the issuance number of each half of the year stabilized at 309. However, with the promulgation of *Further Guideline on CDM Project* in late 2010, the issuance increased again as evidenced by 734 in the first half of 2011 or 150 $MtCO_2e$, 2-3 times of previous issuances.

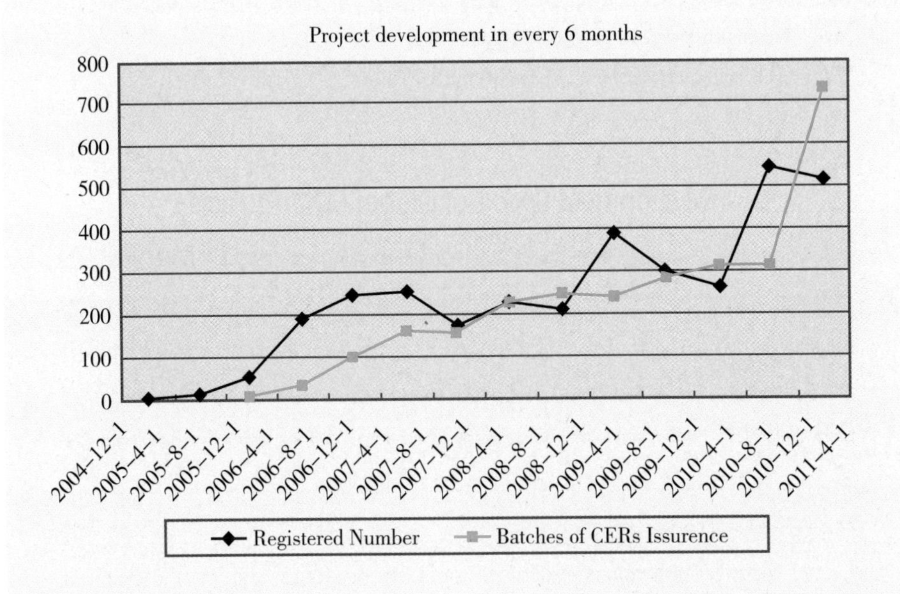

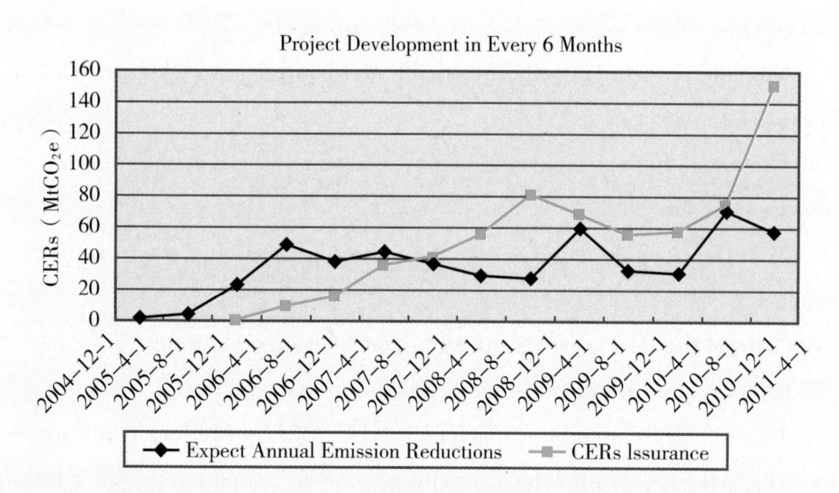

Diagram 3-4 Development trend of global CDM projects

Source: Data collected from the website of CDM EB, UNFCCC.

3.2 Development of China's CDM Projects

The development of China's CDM projects is among the fastest and most mature in the world. This is mainly due to government's emphasis on CDM projects development and deployment and stable national situation.

3.2.1 China's CDM Project Types

China CDM projects were mainly distributed in the following 11 areas, including:

(1) New energy and renewable energy: wind power, solar power, hydropower, geothermal power, tidal power, etc;

(2) Energy conservation and energy efficiency improvement: energy-saving technologies and utilization of waste heat and pressure, etc;

(3) Methane recycling: coalbed methane and landfill gas recycling, household methane use, etc;

(4) Alternative fuel: power generation by using natural gas instead of coal, boiler coal replacement, etc;

(5) Materials replacement: cement production by using waste rather than limestone,etc;

(6) Garbage disposal: waste incineration, refuse composting, etc;

(7) Resources recycling: recycling of industrial waste gas, etc;

(8) HFC-23 decomposition: to change HFC-23(a by-product in the process of producing HCFC-22) into the gas with no or less greenhouse effect through incineration or catalysis;

(9) N_2O decomposition: to change N_2O (a by-product in the process of producing nitric acid and adipic acid) into a gas with no or less greenhouse effect through incineration or catalysis;

(10) SF_6 recycling;

(11) Afforestation and reforestation.

3.2.2 Projects with National Approval

Since China's first CDM project was approved on 25th January 2005, 3,104 projects in China have been approved by DNA by the end of 30th June 2011, with expected annual emission reductions of 490 $MtCO_2e$.

Geographical distribution of projects. These projects were evenly distributed in all provinces (municipalities and autonomous regions) except Tibet, without any province playing a dominant role. According to the statistics of project number, the top six provinces were Yunnan, Sichuan, Inner Mongolia, Gansu, Hunan and Shandong. They accounted for 43.7% of total projects. Since the industries on which each province focused were different (the eastern area focused on industrial projects, while the southwest area mainly concentrated on hydropower, wind power and other new energy projects), the distribution of expected annual emission reductions was different from that of project number. If measured by the expected annual emission reductions, the top six would be Sichuan, Jiangsu, Inner Mongolia, Zhejiang, Shanxi and Guangdong. They took up 45.7% of total reductions. The distribution of projects with national approval was shown in table 3-5 and table 3-6.

Table 3-5 Distribution of projects with national approval among provinces

Province	Number of Projects	Percentage (%)
Yunnan	327	10.5
Sichuan	290	9.3
Inner Mongolia	253	8.2
Gansu	165	5.3
Hunan	164	5.3
Shandong	157	5.1
Others	1,748	56.3
Aggregate	3,104	100.0

Source: http://cdm.ccchina.gov.cn/web/index.asp.

Table 3-6 Distribution of expected annual emission reductions among provinces

Province	Expected Annual Emission Reductions ($MtCO_2e$)	Percentage (%)
Sichuan	51.71	9.9
Jiangsu	41.51	7.9
Inner Mongolia	38.58	7.4
Zhejiang	37.78	7.2
Shanxi	35.45	6.8
Shandong	34.74	6.6
Others	284.83	54.3
Aggregate	524.60	100.0

Note: 1 $MtCO_2e$=1,000,000 tCO_2e.

Source: http://cdm.ccchina.gov.cn/web/index.asp.

Distribution among industries. According to the statistics of project number, these projects were mainly in new energy and renewable energy, energy conservation and energy efficiency improvement and methane recycling industries. The projects in these industries accounted for 95.6% of total projects. However, if measured by expected annual emission reductions, the projects were mainly in new energy and renewable energy, energy conservation and energy efficiency improvement, HFC-23 decomposition, methane recycling, N_2O decomposition, and alternative fuel. Projects in these industries took up 98.7% of total emission reductions. It is worth noting that though there were only 11 HFC-23 decomposition projects and 32 N_2O decomposition projects, still their expected annual emission reductions ranked 3rd and 5th respectively since the two gases have high Global Warming Potential (11,700 and 310 times that of CO_2 respectively) and had enormous reductions. The distribution of projects among industries was shown in table 3–7 and table 3–8.

Table 3–7 Distribution of project number among industries

Industry	Number of Projects	Percentage (%)
New energy and renewable energy	2,229	71.8
Energy conservation and energy efficiency improvement	521	16.8
Methane recycling	216	7.0
Alternative fuel	42	1.4
Materials replacement	31	1.0
N_2O decomposition	32	1.0
HFC-23 decomposition	11	0.4
Garbage disposal	9	0.3
Resources recycling	7	0.2
Afforestation and reforestation	4	0.1
SF_6 recycling	2	0.1
Aggregate	3,104	100.0

Source: http://cdm.ccchina.gov.cn/web/index.asp.

Table 3–8 Distribution of expected annual emission reductions among industries

Industry	Expected Annual Emission Reductions ($MtCO_2e$)	Percentage (%)
New energy and renewable energy	277.30	52.9
Energy conservation and energy efficiency improvement	83.65	15.9
HFC-23 decomposition	66.90	12.8
Methane recycling	52.01	9.9
N_2O decomposition	24.52	4.6
Alternative fuel	13.53	2.6
Materials replacement	4.89	0.9
Garbage disposal	1.07	0.2
SF_6 recycling	0.32	0.1
Resources recycling	0.30	0.1
Afforestation and reforestation	0.12	0.0
Aggregate	524.60	100.0

Note: 1 $MtCO_2e$=1,000,000 tCO_2e.

Source: http://cdm.ccchina.gov.cn/web/index.asp.

3.2.3 Project Registration

Since China's first CDM project got registered on 26th June 2005, 1,516 projects in China have been registered by CDM EB by the end of 30th June 2011, accounting for 45.0% of the total registered by UN (3,368); the expected annual emission reductions is 319 $MtCO_2e$, about 63.5% of the world total (507 $MtCO_2e$), more than any other countries.

Geographical Distribution. According to the statistics of project number, China's registered projects were distributed in all provinces (municipalities and autonomous regions) except Tibet Autonomous Region. The top six provinces and autonomous regions, namely Yunnan, Inner Mongolia, Sichuan, Gansu, Hunan and Shandong, accounted for 47.8% of total projects. If measured by expected annual emission reductions, the top six provinces and autonomous regions were Zhejiang, Jiangsu, Sichuan, Inner Mongolia, Shandong and Shanxi. The ranking was rather different from that of project number. Among the six provinces, Inner Mongolia's expected annual emission reductions ranked forth due to its large number of registered projects, the other five provinces were among the top because their HFC-23 and N_2O decomposition projects could bring large emission reductions. The six provinces' annual emission reductions took up 54.7% of total reductions. The distribution of registered projects among provinces was shown in table 3–9 and table 3–10.

Table 3–9 Distribution of registered project number among provinces and autonomous regions

Region	Number of Projects	Percentage (%)
Yunnan	171	11.3
Inner Mongolia	162	10.7
Sichuan	139	9.2
Gansu	106	7.0
Hunan	86	5.7
Shandong	60	4.0
Hebei	58	3.8
Others	734	48.4
Aggregate	1,516	100.0

Source: Data collected from the website of CDM EB, UNFCCC.

Table 3–10 Distribution of expected annual emission reductions among provinces and autonomous regions

Region	Expected Annual Emission Reductions ($MtCO_2e$)	Percentage (%)
Zhejiang	32.67	10.3
Jiangsu	31.45	9.9
Sichuan	27.72	8.7
Inner Mongolia	25.48	8.0
Shandong	22.83	7.2
Shanxi	20.11	6.3
Others	158.41	49.7
Aggregate	318.67	100.0

Note: 1 $MtCO_2e$=1,000,000 tCO_2e.

Source: Data collected from the website of CDM EB, UNFCCC.

Distribution among industries. If measured by the project number, the registered projects were mainly in new energy and renewable energy, energy conservation and efficiency improvement and methane recycling industries. They took up 94.7% of total registered projects. In terms of expected annual emission reductions, the projects were mainly in new energy and renewable energy, HFC-23 decomposition, alternative fuel, methane recycling, energy conservation and energy efficiency improvement and N_2O decomposition industries. They accounted for 99.5% of total emission reductions. The distribution of registered projects among industries was shown in table 3–11 and table 3–12.

Table 3–11 Distribution of registered project among industries

Industry	Number of Projects	Percentage (%)
New energy and renewable energy	1,220	80.5
Energy conservation and efficiency improvement	109	7.2
Methane recycling	107	7.1
Others	80	5.3
Aggregate	1,516	100.0

Source: Data collected from the website of CDM EB, UNFCCC.

Table 3–12 Distribution of expected annual emission reductions among industries

Industry	Expected Annual Emission Reductions ($MtCO_2e$)	Percentage (%)
New energy and renewable energy	149.12	46.8
HFC-23 decomposition	65.65	20.6
Alternative fuel	30.03	9.4
Methane recycling	28.36	8.9
Energy conservation and energy efficiency improvement	22.76	7.1

Renewal table

Industry	Expected Annual Emission Reductions ($MtCO_2e$)	Percentage (%)
N_2O decomposition	21.02	6.6
Others	1.72	0.5
Aggregate	318.67	100.0

Note: 1 $MtCO_2e$=1,000,000 tCO_2e.

Source: Data collected from the website of CDM EB, UNFCCC.

3.2.4 Issuance of CERs

Since China's first CERs was issued on 3rd July 2006, 368 $MtCO_2e$ of CERs (1125 batches) from 489 projects in China were issued by the end of 30th June 2011, accounting for 56.9% of the world total (647 $MtCO_2e$), far above India (15.1%) and No.1 in the world.

Geographical distribution. CERs were issued in all provinces (municipalities and autonomous regions) except Tibet Autonomous Region. The top six provinces and autonomous regions were Zhejiang, Jiangsu, Shandong, Liaoning, Sichuan and Inner Mongolia. They all have HFC-23 and N_2O decomposition projects characterized by large reductions, and accounted for 81.3% of total CERs issuance. The CERs issuance distribution was shown in table 3–13.

Table 3–13 CERs issuance distribution among provinces and autonomous regions

Region	Number of Projects	Issuance Batch	Amount of Issuance ($MtCO_2e$)	Percentage (%)
Zhejiang	16	75	88.93	24.2
Jiangsu	19	82	88.57	24.1
Shandong	18	71	59.35	16.1
Liaoning	17	46	34.39	9.3
Sichuan	37	77	15.62	4.2

Renewal table

Region	Number of Projects	Issuance Batch	Amount of Issuance ($MtCO_2e$)	Percentage (%)
Inner Mongolia	50	99	12.29	3.3
Others	332	675	68.91	18.7
Aggregate	489	1,125	368.07	100

Note: 1 $MtCO_2e$=1,000,000 tCO_2e.

Source: Data collected from the website of CDM EB, UNFCCC.

Distribution among industries. CERs were mainly in HFC-23 decomposition, new energy and renewable energy and N_2O decomposition industries. They took up 90.7% of total CERs. HFC-23 decomposition projects accounted for 62.6 percent, much higher than other projects. This was mainly due to: (1) the HFC-23 and N_2O decomposition projects had large scale and were implemented earlier; (2) although the scale of new energy and renewable energy project was small, still the quantity was huge. The distribution of CERs among industries was shown in table 3–14.

Table 3–14 Distribution of CERs among industries

Industry	Number of Projects	Issuance Batch	Issuance Amount ($MtCO_2e$)	Percentage (%)
HFC-23 decomposition	11	138	230.41	62.6
New energy and renewable energy	377	738	59.77	16.2
N_2O decomposition	12	48	43.51	11.8
Alternative fuel	16	47	14.52	3.9
Energy conservation and energy efficiency improvement	45	90	12.93	3.5
Methane recycling	28	64	6.92	1.9
Aggregate	489	1,125	368.07	100.0

Note: 1 $MtCO_2e$=1,000,000 tCO_2e.

Source: Data collected from the website of CDM EB, UNFCCC.

The general situation of China's CDM projects was shown in table 3–15 and table 3–16.

Table 3–15 Geographic distribution of China CDM projects

Region	National Approval		Registration		CER Issuance		
	Number	Expected Annual Emission Reductions ($MtCO_2e$)	Number	Expected Annual Emission Reductions ($MtCO_2e$)	Number	Batch	Issuance Amount ($MtCO_2e$)
Yunnan	327	33.5	171	18.6	47	75	4.86
Sichuan	290	51.7	139	27.7	37	77	15.62
Inner Mongolia	253	38.6	162	25.5	50	99	12.29
Gansu	165	23.9	106	15.2	25	50	5.24
Hunan	164	15.0	86	8.5	35	64	6.04
Shandong	157	34.7	60	22.8	18	71	59.35
Hebei	147	19.0	58	8.2	21	59	5.40
Shanxi	121	35.4	53	20.1	12	36	6.82
Henan	117	19.0	34	9.5	14	35	11.38
Zhejiang	105	37.8	32	32.7	16	75	88.93
Hubei	99	9.9	50	6.7	13	22	1.66
Jiangsu	97	41.5	40	31.4	19	82	88.57
Liaoning	97	27.7	48	19.8	17	46	34.39
Guizhou	91	9.0	44	2.6	18	29	1.38
Jilin	91	12.9	40	5.7	11	31	2.61
Guangdong	84	11.3	42	9.6	20	36	2.58
Fujian	83	10.6	50	9.3	21	32	2.61
Guangxi	81	11.7	42	4.5	11	15	0.96
Heilongjiang	73	15.7	23	4.8	11	25	1.40
Shaanxi	72	8.5	35	4.4	4	4	0.17
Xinjiang	69	11.2	43	7.2	12	32	2.72
Anhui	58	8.6	23	3.4	12	21	2.25
Chongqing	58	10.2	35	6.5	9	28	1.76

Renewal table

Region	National Approval		Registration		CER Issuance		
	Number	Expected Annual Emission Reductions ($MtCO_2e$)	Number	Expected Annual Emission Reductions ($MtCO_2e$)	Number	Batch	Issuance Amount ($MtCO_2e$)
Jiangxi	56	5.0	31	2.3	10	15	0.88
Ningxia	55	5.4	27	3.2	10	30	2.97
Qinghai	28	2.1	14	1.3	5	13	1.11
Shanghai	19	6.9	6	3.4	1	1	0.03
Hainan	18	0.9	10	0.6	4	8	0.22
Beijing	16	4.7	9	2.8	5	13	3.82
Tianjin	13	2.1	3	0.3	1	1	0.03
Aggregate	3,104	524.6	1,516	318.7	489	1,125	368.10

Note: The statistics were by the end of 30 June 2011.

Source: Data collected from the website of CDM EB, UNFCCC.

Table 3–16 Distribution of projects among industries

Project	National Approval		Registration		CERs Issuance		
	Number of Projects	Expected Annual Emission Reductions ($MtCO_2e$)	Number of Projects	Expected Annual Emission Reductions ($MtCO_2e$)	Number of Projects	Batch	Expected Annual Emission Reductions ($MtCO_2e$)
New energy and renewable energy	2,229	277.3	1,220	149.1	377	738	59.8
Energy conservation and energy efficiency improvement	521	83.7	109	22.8	45	90	12.9
Methane recycling	216	52.0	107	28.4	28	64	6.9
Alternative fuel	42	13.5	29	30.0	16	47	14.5
N_2O decomposition	32	24.6	27	21.0	12	48	43.5

Renewal table

Project	National Approval		Registration		CERs Issuance		
	Number of Projects	Expected Annual Emission Reductions ($MtCO_2e$)	Number of Projects	Expected Annual Emission Reductions ($MtCO_2e$)	Number of Projects	Batch	Expected Annual Emission Reductions ($MtCO_2e$)
Materials replacement	31	4.9	5	1.2	0	0	0.0
HFC-23 decomposition	11	66.8	11	65.7	11	138	230.4
Garbage treatment	9	1.1	4	0.2			0.0
Resources recycling	7	0.3	0	0.0	0	0	0.0
Afforestation and reforestation	4	0.1	3	0.1	0	0	0.0
SF_6 recycling	2	0.3	1	0.2			0.0
Aggregate	3,104	524.6	1,516	318.7	489	1,125	368.1

Note: The statistics were by the end of 30 June, 2011.

Source: Data collected from the website of CDM EB, UNFCCC.

3.2.5 Experiences in Developing and Implementing CDM Projects

China boasts the largest number of CDM projects with high quality in the world. Moreover, the projects are developing fast. This is mainly due to[1]:

3.2.5.1 Government Paid Close Attention and Provided Substantial Support to CDM Work

(1) Establishment of regulatory agency. China set up a special government authority—CDM project review board at the start of CDM implementation, composed of seven ministries, namely National Development Referm Commission (NDRC),

1 China CDM Fund Management Center. *Climate Change Financing* [M]. Beijing: Economic Science Press, 2011: 274–276.

Ministry of Foreign Affairs, Ministry of Science and Technology, Ministry of Finance, Ministry of Environmental Protection, Ministry of Agriculture and China Meteorological Administration. The Board is in charge of review and approval of CDM projects. Meanwhile, NDRC set up a climate change office (climate change department), responsible for project development and implementation, project review and approval, and also serve as DNA to communicate with CDM Executive Board of United Nations.

(2) Promulgation of regulations and standards. To facilitate project implementation, on 30th June 2004, China issued *the Interim Measures for Management of CDM Project Operation,* to standardize project development procedure. After one year's practices and accomodating new situations and problems, China promulgated the *Measures for Management of CDM Project Operation* and made a further revision in August 3rd, 2011. These regulations have standardized project development and approval procedure and ensured open, transparent, fair and orderly implementation.

(3) Carry out nationwide capacity building activities. Seeing that CDM project is a new international innovation, the government has vigorously conducted nationwide capacity building since the beginning of CDM operation by holding workshops, seminars, and other experience exchanges on the area to people from governments, enterprises and other related institutions. This has widely communicated the concept of CDM, and helped many enterprises and institutions grasp key expertise and technologies, and grow stronger through practices.

(4) Enterprises and national government share CERs revenue. To make CDM carbon trade better support China tackle climate change, and considering GHG emission reductions are public resources, China stipulated that national government and enterprises share revenue from the trade, and established China Clean Development Mechanism Fund (China CDM Fund) and its management center to collect, manage, and utilize the national revenue. The management center uses the fund specifically to support national climate change actions, including CDM work. The revenue sharing arrangement elevates CDM cooperation to country-level, and the gathering of national revenue helps strategically promote low carbon economy and sustainable development.

3.2.5.2 China's Economic Development Offers Opportunities for CDM Project Implementation

In recent years, Chinese government highly prioritizes building an environmentally-friendly and resource-efficient society, and has made great effort to promote energy conservation and emission reduction, and facilitates economic restructuring. On one hand, the country is phasing out backward capacity, improving industrial technology and reducing energy consumption; on the other, China actively supports development of new energy and renewable energy. These are great project resources for CDM development in China. Meanwhile, CDM projects also push forward enterprises' technology upgrades and industrial transition. Also important is the great variety of industrial projects in China, including large CDM projects like HFC-23 and N_2O decomposition.

3.2.5.3 Abundant Natural Resources in China Cultivates Diversified CDM Projects

China has a vast territory and is rich in wind power, hydro power, solar energy, geothermal energy, tidal energy, etc, forming a wide and diversified base for developing CDM projects.

3.2.5.4 Consultancies Are Important Engine for CDM Project Development

Due to the complex CDM rules and procedures with English as the official language and international trade laws as reference, many project developers find it is hard to develop and implement projects by themselves. Realizing this problem at the very beginning, China has cultivated many professional CDM consultancies. They were driven by market to seek potential projects, promote CDM to possible developers and provide one-stop service, which greatly helped universalize CDM. According to initial statistics, there have been over 100 CDM consultancies in China, most of which are small and medium enterprises (SMES). This is a great support to SMEs and an impetus for economic restructuring.

The Chinese government also develops domestic independent third party verification agencies. Currently, China has four such agencies approved by CDM Executive Board of UN. This has facilitated project progress by removing language and cultural obstacles from foreign agencies doing the verification.

Chapter Ⅳ

Major Roles and Problems of CDM

4.1 Major Roles of CDM

CDM was widely viewed as a win-win mechanism. Developed countries could fulfill their commitments of Kyoto Protocol in a cost-effective way, and transfer technologies, products and services to developing countries. Developing countries could obtain CDM concept, technologies and emission reduction funds, which further promoted sustainable development and environmental protection. Through several years' practices, CDM projects made great contributions to the climate change in the world, reflected in the following aspects[1].

4.1.1 Helped Developed Countries Reduce Emission Reduction Cost

CDM was the only effective mechanism at present that united developed and developing countries to address climate change. It enabled developed countries to fulfill commitments of emission reduction in a cost-effective way. According to consultancy McKinsey, the cost of reducing 1 tCO_2e in developed countries was as high as 70–100 dollars[2]. However, the cost of buying 1 tCO_2e from developing countries' CDM primary market was only 5–15 dollars. Through the operation of CDM projects, developing countries helped developed

1 Xie Fei, Meng Xiangming, Hu Ye. CDM: Leveraging Low Carbon Economy in Developing Countries[N]. *China Finance and Economic News*, 21st Jan, 2010: 4.

2 Impetuses and Distortions in China CDM. http://news.sina.com.cn/c/sd/2009-12-23/140519322122.shtml.

countries reduce 11–65 billion dollars of emission reduction fund only in 2010, with the accumulated fund being 42–244 billion dollars. To some extent, China has saved developed countries a significant amount of compliance cost through CDM projects.

4.1.2 Provided Financial Support for Low Carbon Development of Developing Countries

CERs transactions directly brought 1.2 billion dollars to developing countries in 2010; starting from the implementation of CDM projects, the accumulated fund has exceeded 5 billion dollars. China accounted for 40 percent with its 2 billion dollars. Meanwhile, in the process of developing CDM projects, the indirectly leveraged capital was worth tens of billions of dollars. The financial support, to some extent, promoted the low carbon development of developing countries. Particularly during the financial crisis, direct and indirect financial support to developing countries provided timely help, which relieve financial stress of project enterprises and kept their business going.

4.1.3 Provided a New Concept on Sustainable Development for Developing Countries

CDM marketized effectively environmental protection activities, encourage more players to be engaged, and provided new thoughts and practices for developing countries on utilizing market to solve environmental issues. At the same time, developing countries gained advanced corporate philosophy and management experience, and improved scientific, standardized and sophisticated management of enterprises through implementing CDM projects and communicating with CDM EB and DOE. Through 5 years' operation in China, many companies understand further energy conservation and emission reduction and low carbon development as well as how developed countries use market mechanism (carbon market) in this area. Until now, China has designated 5 provinces and 8 municipalities as pilot regions of low carbon development[1], and put forward the idea of gradually establishing carbon trade market during the 12th Five-year Plan period[2]. CDM undoubtedly paves a foundation for all these activities.

4.1.4 Developed a Globalized Work Team of Environmental Protection for Developing Countries

Project owners, consultancies and DOEs in developing countries work according to international rules, which can unwittingly enhance the capacity of staff in developing countries. This is of great importance for developing world to carry out other environmental activities. This was quite evident in China's CDM projects. Many enterprises reported that the implementation of CDM projects broadened the vision of their working staff, and improved their capacity and expertise.

CDM projects have enhanced the capacity of a group of domestic certification bodies in the process. So far four certification bodies have become DOEs, namely China Environmental United Certification Center CO., Ltd. (CEC), China Quality Certification Center (CQC), and China Classification Society Certification Company (CCSC) and CEPREI Certification body (CEPREI). These entities not only facilite the operation of CDM projects, but also improve the quantification of domestic energy conservation and emission reduction, making it closer to international standards.

4.2 Major Problems of CDM

4.2.1 Unbalanced Development in the World

According to each country's registration number, expected annual emission reductions and CERs issuance, China, India and Brazil jointly monopolized the

1 NDRC: Circular on Low Carbon Pilot Programs in Certain Provinces and Cities. http://www.sdpc.gov.cn/zcfb/zcfbtz/2010tz/t20100810_365264.htm.

2 The 12th Five–Year Plan of PRC. http://news.xinhuanet.com/politics/2011–03/16/c_121193916_12.htm.

market with a share of 71%. This testified that CDM, as an UN-led mechanism, could not benefit all developing countries, and promote all their sustainable development, which made China, India and Brazil the target of public criticism. This harmed CDM development, discouraged other developing countries, thus affecting global response to climate change. Major causes of this problem are: (1) The long operational cycle of CDM projects demands high level of stability of enterprises, so international buyers were willing to choose capable countries with political stability. (2) Compared with large developing countries, small developing countries and island nations find it difficult to develop large CDM projects because of fewer large enterprises and low GHG emission base. From the perspective of reducing the transaction cost, international buyers are more willing to purchase large-scale CDM projects. (3) CDM projects rules were complex, which made some countries less able to carry out CDM[1].

4.2.2 CDM Development Far from the Needs for Addressing Climate Change of the World

Global CDM development at present falls far behind the needs for addressing climate change of the world due to many restraints. For example, by 30th June 2011, only 1,088 projects in the world acquired CERs' issuance, accounting for 32.3% of global registered projects (3,368). Among 3,104 projects approved by China's authority, 1,516 and 489 projects registered successfully and gained CER issuance respectively, taking up 48.8% and 15.8% of the total approved projects. The restraints included: (1) CDM procedures were complex, and the methodology was applied narrowly and updated quickly. (2) The CDM EB had low work efficiency, leading to the backlog of registration and CERs issuance. (3) The DOEs' ability was limited. By 30th June 2011, only 38 consultancies became qualified DOE in the world. Since EB approved 11 consultancies in 2003 when CDM projects just started operation, they have been developing very slowly. (4) The decision-making process of CDM was not transparent. EB reserved the discretion to decide many important rules. For example, in the 2010 *CDM Validation and Verification Manual*, many

1 Meng Xiangming, Feng Chao, Xie Fei. Global CDM Market Development and Challenges[J]. *Economic Research*, 2009, 2217(17) :5–9.

parameters and operation details were very vague[1].

4.2.3 CDM Could Only Be Regarded as Beneficial Supplement to Developing Countries Work on Addressing Climate Change

CDM primary market in the world had small scale and low transaction price, so it brought limited fund to developing countries. According to the *State and Trends of the Carbon Market 2011* published by World Bank[2], with the stabilization of global carbon market since it took off in 2005, CDM primary market (which is directly relevant to developing countries) value has brought 26.5 billion dollars to the developing world (only 5% of global carbon market), with its share as a part of global carbon market gradually reducing from 23.6% in 2005 to less than 1.1 percent in 2010. Moreover, due to the policy uncertainties of second commitment period of Kyoto Protocol and EU bashing at CDM, the CDM primary market value decreased at a double digit rate after reaching 7.4 billion dollars in 2007. In 2010, the transaction value was only 1.5 billion dollars, down by 44.4% and lower than the 2.6 billion dollars of 2005 (The global carbon market development since 2005 was shown in table 4–1). We can see that, without reform, CDM can not become an important source of capital for developing countries as expected by developed countries[3].

Table 4–1 2005–2010 global carbon market trading value unit:Billon dollars

Year	CDM Primary Market	Global Carbon Market	Percentage (%)
2005	26	110	23.6
2006	58	312	18.6
2007	74	630	11.7
2008	65	1,351	4.8

1 Xie Fei, Meng Xiangming, Liu Miao. Global Carbon Market: Looking to Break a Bottleneck[N]. *China Finance and Economic News*, 24th, Jun, 2010: 4.

2 Carbon Finance at World Bank. *State and Trends of the Carbon Market*, 2011: 9.

3 Xie Fei, Xu Mingzhu, Meng Xiangming. EU's Latest Cliamte Change Strategy[N]. *China Finance and Economic News*, 8th April, 2010: 4.

Renewal table

Year	CDM Primary Market	Global Carbon Market	Percentage（%）
2009	27	1,437	1.9
2010	15	1,419	1.1
合计	265	5,259	5.0

Source：Carbon Finance at World Bank. *State and Trends of the Carbon Market 2011.*

According to the technical report published by UNFCCC secretariat in 2007, the funds demands of developing countries would reach 120–163 billion dollars by 2030[1]. The *Copenhagen Accord* stipulated that developed countries committed to provide 30 billion dollars each year from 2010 to 2012, and jointly mobilize 100 billion dollars a year from 2013 to 2020[2]. UN Secretary General's High-Level Advisory Group on Climate Change Finance suggested that 100 billion dollars could be sourced from carbon market.

Compare with the funds demands and supplies, we can see that, if CDM is not reformed, the funds that developing countries could acquire from carbon market would be limited, far from satisfying the demands of addressing climate change.

Some Key Index of CDM Projects

The duration from registration to CERs issuance and the duration from the end of a crediting period to the next CERs issuance are two important factors to measure the speed of projects.

（1）the duration from registration to CERs issuance. By the end of 30th June 2011, 1,088 projects have received CERs issuance. By analyzing these projects, we found the average time taken to receive issuance

1 China CDM Fund Management center. *Climate Change Financing* [M]. Beijing: Economic Science Press, 2011: 56.

2 UNFCCC, Decisions adopted by the Conference of the Pargties. http://unfccc.int/resource/docs/2009/cop15/eng/11a01.pdf, P6.

was 570 days. With the registration application number increasing, the duration prolonged. In the latter half of 2010 and first half of 2011, UN EB sped up registration. By the end of April 2011, those backlog projects were issued, with the duration longer than the average. However, since May 2011, the duration has been reducing with the final handling of backlog projects. These are shown in Diagram 4-1.

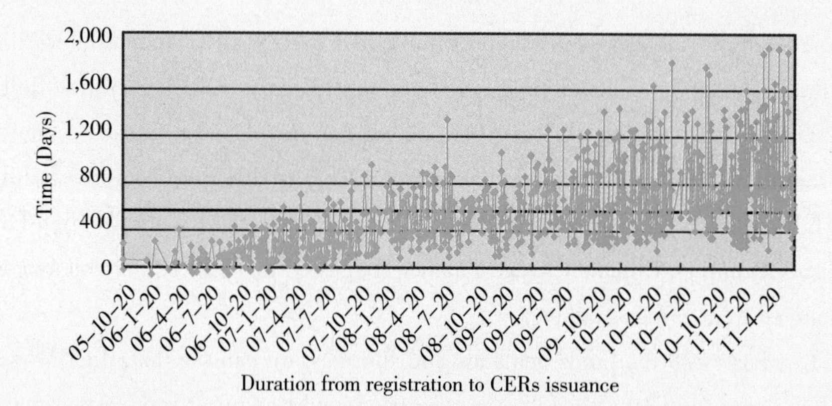

Diagram 4-1 Duration from registration to CERs issuance

Source: Data collected from the website of CDM EB, UNFCCC.

(2) the duration from the end of a crediting period to the CERs issuance. CERs issuance duration referred to the duration from the end of a crediting period to the next issuance of CERs. By analyzing 2,789 CERs issuance in the world, the average time was 302 days. However, the average time increased later with rising issuance applications. In the latter half of 2010 and the first half of 2011, the backlog applications were dealt with altogether. Since June 2011, the average time was reduced to 363 days. The issuance with average time of 101-400 days accounted for 69.8% of total issuances. The CERs issuance trend was shown in diagram 4-2 and diagram 4-3.

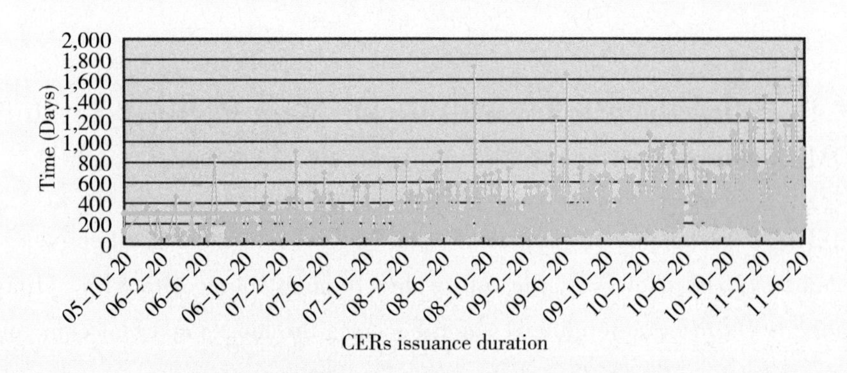

Diagram 4–2 CERs issuance duration

Source: Data collected from the website of CDM EB, UNFCCC.

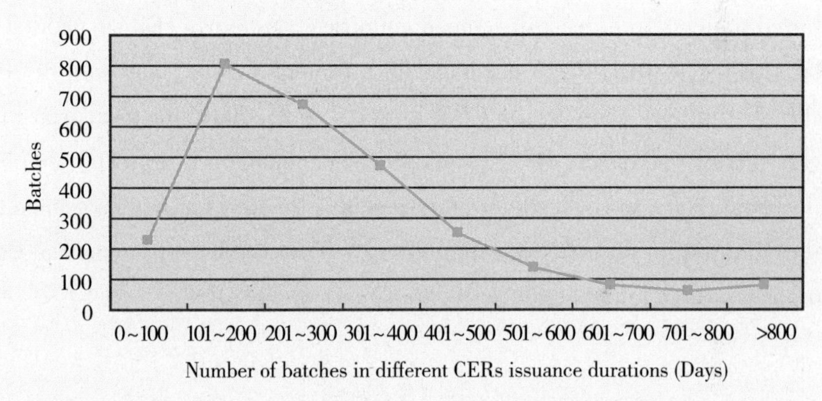

Diagram 4–3 Number of batches in different CERs issuance durations

Source: Data collected from the website of CDM EB, UNFCCC.

Through the analysis of global CDM projects development, we found that EB sped up the registration in 2010, so the new registration number and CERs issuance increased dramatically. However, since the whole process involved many links, such as certification by DOE, submission for CERs issuance, etc, the duration was increasing in fact. By the end of 2010, the duration from registration to CERs issuance was as long as 2 years, while the CERs issuance duration was 1 year.

4.2.4 Policy Uncertainty Will Be the Main Factor Restraining CDM Development

The first commitment period of Kyoto Protocol is coming to an end on 31st, December 2012. However, the future development path of tackling climate change is yet to be determined, which becomes the most important constraint. In the Directive 2003/87/EC of the European Parliament and of the Council of 23rd April 2009 (revised on 25th June 2009), the EU emission trading system, starting from the third phase, would only accept emission reductions of newly registered projects from least developed countries[1]. Moreover, EB suspended the CERs issuance of all HFC-23 decomposition projects in the latter half of 2010. Western institutions' criticism on CDM projects, particularly the industrial waste projects (mainly including HFC-23 and N_2O decomposition projects) shattered people's confidence in CDM. Before the Cancun Climate Change Conference, EU announced a resolution to forbid international carbon credit from industrial waste gas projects, which undoubtedly made the situation even worse.

1 Directive 2003/87/EC of the European Parliament and of the Council of 23 April 2009, http://eur-lex.europa.eu/LexUriServ/LexUriServ.do?uri=OJ:L:2009:140:0063:0087:EN:PDF, P24.

Chapter V

Prospects of CDM

Policy uncertainties in the future and developed countries' boycott to CDM projects in developing countries has made CDM come into a deadlock. Many countries including China balked at CDM projects after 2012. However, in the Directive 2003/87/EC of the European Parliament and of the Council of 23rd April 2009 (revised on 25th June, 2009), EU stipulated that, before any climate change protocol was reached, EU Emission trading system (EU ETS) would continue to accept CERs generated before 2012 from compliance projects under EU ETS from 2008 to 2012, as well as CERs generated after 2013 from compliance projects registered before 31st December, 2012[1]. This demonstrated that EU continued to utilize the emission reductions of CDM projects after 2012.

In the *Cancun Agreement* adopted at the 2010 Cancun climate change conference, all parties agreed that the first and second commitment period of Kyoto Protocol should be connected ceaselessly, and promised to cultivate more market instruments based on current market mechanism[2]. At this conference, they also adopted *Further guidance relating to the clean development mechanism*, and put forward explicit requests on CDM reform. All these augur a bright future[3].

The package of protocols passed at the COP 17 in Durban in 2011 also

---

1 UNFCCC, Decisions adopted by the Conference of the Pargties. http://unfccc.int/resource/docs/2009/cop15/eng/11a01.pdf, P6.

2 UNFCCC, Report of the Conference of the Parties serving as the meeting of the Parties to the Kyoto Protocol on its sixth session, held in Cancun from 29 November to 10 December 2010, P4.

3 Meng Xiangming, Li Chunyi, Xie Fei. Carbon market and CDM will not disappear after 2012[N]. *China Energy News*, 24th, Jan, 2011 (P6).

stipulated that the second commitment period of Kyoto Protocol starts from 1st January 2013 and ends at 31st December 2017 or 31st December 2020. This is in line with the "dual track"negotiation approach held by developing countries, maintained second commitment of Kyoto Protocol, and somewhat saved CDM, a most successful mechanism widely used in China and contributes a lot to China's sustainable development[1].

CDM in the second commitment period of Kyoto Protocol is still uncertain on many technical issues, and EU will further restrict CDM from many emerging economies including China. However, the fact that CDM will continue to exist boosts the confidence of Chinese project owners and consultancies. Even if CDM disappeared one day, various carbon markets and related mechanisms would exist because their effectiveness has been recognized by all parties and proved by many facts.

1 Europe has a plan to keep CDM if Kyoto Protocol expires. http://www.rechargenews.com/hardcopy/article 274560.ece.

Part Ⅱ

China Clean Development Mechanism Fund

Chapter VI

The Origin of China Clean Development Mechanism Fund

6.1 National Revenue from CDM Projects

CDM is project-based cooperation, in that it is cooperation between the emission reduction buyers of developed countries and the CDM project companies of developing countries, with the former providing funding, the latter producing CERs. However, the government of developing countries, the one short of funding for coping with climate change, has no way to receive any revenue in the process of developing and implementing CDM projects. This is a weakness of CDM. Undoubtedly it will be more effective if the government of developing countries can obtain certain amount of revenue from CDM cooperation to support its efforts of addressing climate change at national level.

Chinese government has realized from the beginning of its CDM endeavor that emission reduction capacity of GHG, like mineral and other environment resources, is a kind of public resource that should belong to the government; therefore, revenues from emission reduction transactions should be shared by the government and the project owners. The part of revenue taken by the government is called national revenue. The fact that China is rich in HFC-23 and N_2O projects that can produce massive reduction further contributes to the proposal of national revenue.

Both HFC-23 and N_2O are waste gases generated from chemical productions and were discharged without any treatment before the introduction of CDM projects. Global Warming Potential (GWP) of HFC-23 and N_2O are 11,700 and 310 times that of CO_2 respectively, which will be a huge influence on global warming. CDM projects on HFC-23 and N_2O decomposition have one thing in common, that is both can produce massive reduction and cost less compared with wind power, hydropower and energy efficiency projects. Those two types of projects mainly concentrate in China, India and several other countries. The international community has been concerned that the high profitability from HFC-23 and N_2O decomposition projects may incentivize the companies to produce on mal-purpose the main products just for getting CDM revenues regardless of the market demand. In response to this, Chinese government, as a responsible player, decides to levy a higher percentage of national revenue on those two types of projects.

Therefore, China released the *Interim Measures for Management of Clean Development Mechanism Projects Operation* on 30th June 2004 which stipulated that, "Revenues from the transfer of CERs generated by CDM projects shall be shared by the Government of China and the project company. The allocation ratio will be determined by the Government of China. And before the ratio is specified, the project company is entitled to the full ownership of the revenues." In October 2005, when *Measures for Management of Clean Development Mechanism Projects Operation* was unveiled, the wording was revised as, "Whereas emission reduction resource is owned by the Government of China and the emission reductions generated by specific CDM project belong to the project owner, revenue from the transfer of CERs shall be owned jointly by the Government of China and the project owner, with the allocation ratio defined as below:

(1) The Government of China takes 65% CER transfer benefit from HFC and PFC projects;

(2) The Government of China takes 30% CER transfer benefit from N_2O project;

(3) The Government of China takes 2% CER transfer benefit from CDM projects in priority areas defined in Article 4 and forestation projects.

The revenue collected from CER transfer benefits of CDM projects will be used in supporting activities on climate change."

On 3 August 2011, in light of the implementation of CDM projects in China, Chinese government revised *Measures for Operation and Management of Clean Development Mechanism Projects,* and released *Measures for Operation and Management of Clean Development Mechanism Projects (Revised Edition)*, which adjusted the ratio of national revenue for different projects as below:

Revenue from the transfer of CERs generated by CDM projects shall be owned jointly by the Government of China and the project institutions, and other institutions and individuals shall not share the revenues. The allocation ratio between the Government of China and the project institutions is specified as below:

(1) The Government of China takes 65% CER transfer benefit from HFC projects;

(2) The Government of China takes 30% CER transfer benefit from N_2O projects related to adipic acid production;

(3) The Government of China takes 10% CER transfer benefit from N_2O projects related to the production of nitric acid and etc;

(4) The Government of China takes 5% of CER transfer benefit from PFC projects;

(5) The Government of China takes 2% of CER transfer benefit from other CDM projects.

The national revenue from CER transfer benefit of CDM projects will be collected by China CDM Fund Management Center in accordance with *Measures for Management of China Clean Development Mechanism Fund,* and be used for supporting activities on combating climate change.

6.2 Initiative for Establishing China CDM Fund

In order to implement *Measures for Operation and Management of Clean Development Mechanism Projects*, and fully play the role of national revenue from CDM projects to effectively support China's efforts in combating climate change, in 2006, Ministry of Finance took the lead and was joined by National Development and Reform Commission, Ministry of Foreign Affairs, Ministry of Science and Technology, Ministry of Environmental Protection, Ministry of Agriculture and China Meteorological Administration in applying to the State Council for the establishment of

China Clean Development Fund (China CDM Fund) and its Management Center. The application was soon approved.

In the application process, China CDM Fund, thoroughly taking reference from management models of mature funds like Social Security Fund, China-Belgium Fund, Global Environment Facility, and the Carbon Trust, meeting the need of rapid development of work on tackling climate change and the new situation of international cooperation on climate change, as well as giving full consideration to the helpful suggestions from international buyers of CERs and organizations like the World Bank and the Asian Development Bank, mapped out its business and operation plan as below:

First, the fund will be raised though diversified sources. The main undertakings of the fund are to collect, manage and utilize the national revenue from CDM projects, and maintain and increase its value as well. Meanwhile, the fund will also actively take in and capitalize on domestic and international resources and explore and develop new models for cooperation so as to win more capital support for China's efforts in combating climate change.

Second, the fund will provide multi-dimensional service for tackling climate change. The fund will use the levied national revenue and funds raised from other sources to serve the overall work of the government on combating climate change, including the following undertakings in this area: (1) support capacity building; (2) support the enhancement of public awareness; (3) promote energy efficiency improvement and energy conservation; (4) promote the development and utilization of renewable energy; (5) support activities with conspicuous mitigation effects; (6) support adaptation activities; (7) support financial activities conducive to the sustainable operation of the fund, etc.

Third, the fund will position itself as a policy institution that integrates government and market functions based on the consideration of reflecting policy direction on the one hand and conducting direct cooperation with business partners in the broad market on the other. Under the guidance and with the support of the government, the fund will carry out extensive domestic and international cooperation, guide and promote emission reductions in the market, and explore market-based long-term mechanism to support the low-carbon development of industries and the development of emerging low-carbon industries.

6.3 Significance of Establishing China CDM Fund

China CDM Fund, as an innovative funding and action mechanism, is of great importance to enhancing the role of CDM in China's efforts of tackling climate change as well as improving China's global image in this regard.

First, the fund upgrades the international cooperation on CDM from project level to national level, which strengthens the role of CDM in China's work on combating climate change. As a national special fund for tackling climate change, the fund is able to serve China's climate efforts at national level. Particularly when development is still the primary task at this stage, the fund can promote systematic and orderly implementation actions on climate change and the enforcement of *China National Climate Change Programme* by pooling part of funds dedicated to supporting the country cope with climate change, like policy research, planning, industrial activities, etc.

Second, as an innovative funding mechanism, the fund can play the role of seed capital and guidance fund to mobilize more social funds to the national efforts of combating climate change. It will support the industrialization, marketization and socialization of China's work on tackling climate change, conserving energy and reducing emissions so as to promote low-carbon development of China.

Third, the fund promotes China's international cooperation on climate change through its Management Center's extensive exchanges and cooperation with implementers of the projects sponsored by the fund as well as international organizations.

Chapter VII

The Establishment of China CDM Fund and Its Governance Structure

7.1 The Establishment of China CDM Fund

7.1.1 The Preparations for Establishing China CDM Fund

The preparation and establishment of the fund have received much attention from senior officials of the government, including Minister of Finance Xie Xuren, Vice Minister of National Development and Reform Commission Xie Zhenhua, Executive Vice Minister of Finance Liao Xiaojun, Vice Minister of Finance Zhu Guangyao, and former head of Discipline Inspection Group of Ministry of Finance He Bangjing, who have shown concern and given advice.

In October 2005, the Ministry of Finance took the lead and was joined by National Development and Reform Commission, Ministry of Foreign Affairs and Ministry of Science and Technology in launching the preparation work for establishing China CDM Fund.

China's move of establishing the fund was warmly received by the international community. Some international development organizations, international financial institutions and government agencies of other countries expressed keen interest and strong will to cooperate with the fund. The World Bank and the Asian Development Bank offered great support in capacity building during the preparation and initial operation period.

In December 2005, Ministry of Finance signed on behalf of the Chinese government the first batch of purchase agreements of CERs valued at USD 500 million covering two HFC-23 decompositon projects with the World Bank (WB). Through consultations, the World Bank agreed to pay USD 6.3 million out of the due national revenue from CER transfer of the two projects in advance as seed capital for the fund. This is the first capital inflow for the fund.

In May 2006, the Asian Development Bank (ADB) aided the fund with a 600,000-dollar technical assistance project to help with the preparation work by designing the governance structure, compiling the basic regulations, improving management capability and etc. In October 2008, ADB funded another 800,000-dollar technical assistance project to help the fund strengthen capacity building in the beginning of its establishment, including formulating rules, regulations and administrative manual, designing and building internet-based management information systems, training staff, carrying out publicity activities at home and abroad and etc, which made great contribution to enhancing the capability of the fund for smooth operation and better domestic and international recognition.

7.1.2 The Establishment of China CDM Fund

Under the joint efforts of several ministries and after a year of preparations, in May 2006, Ministry of Finance (MOF), National Development and Reform Commission (NDRC), Ministry of Foreign Affairs (MOFA), and Ministry of Science and Technology (MOST) submitted a joint application for establishing China CDM Fund and its Management Center to the State Council. The State Council attached great importance to the matter and responded promptly with an official approval signed by principal leaders in August 2006 after hearing opinions and suggestion from relevant ministries.

The establishment of the Fund Management Center received great support from State Commission Office for Public Sector Reform (SCOPSR). In March 2007, the Center registered at State Administration of Public Institution Registration, which made it one of a few newly registered institutions with official staff quota in recent years. In April 2007, under the support of State Administration of Foreign Exchange, China CDM Fund officially opened its exclusive account. In April 2007, the Board of China CDM Fund convened the first meeting, which appointed NDRC as Chair, MOF

Vice Chair of the fund, and officially launched the work on formulating rules and regulations of the fund. In the same month, the fund started its business operation. On 12 April 2007, the fund account witnessed its first national revenue inflow.

7.1.3 The Launch of China CDM Fund

On 9 November 2007, the inauguration ceremony of China CDM Fund and its Management Center was held in Beijing. Minister of Finance Xie Xuren, Vice Minister Li Yong, Vice Minister Zhu Guangyao, Vice Minister of NDRC Xie Zhenhua, Vice Minister of Foreign Affairs Zhang Yesui, ADB President Haruhiko Kuroda attended the ceremony and made important remarks.

Minister Xie Xuren pointed out at the ceremony that the establishment of China CDM Fund was a concrete embodiment of the ministry's efforts in implementing the Scientific Outlook on Development and providing more financial support through innovating financing and investment mechanisms. The Ministry

Finance Minister Xie Xuren and Vice Chairman Xie Zhenhua of NDRC jointly launching China CDM Fund

of Finance would give full play to the role of the fund as an incubator, support and develop more energy conservation and emission reduction projects, and build platform for international cooperation and action in the areas of climate change, energy conservation and emission reduction so as to promote domestic sustainable development and international action on combating climate change. Meanwhile, by combining the fund and fiscal resources, the ministry would guide and attract wide participation from social forces, thus integrate the resources from both public and private sectors to contribute to the building of resource efficient and environmentally friendly society.

Vice Minister of NDRC Xie Zhenhua stressed in his speech that the establishment of China CDM Fund was a milestone in the areas of climate change, energy conservation and emission reduction. As the international community paid closer attention to climate issues and Chinese government attached more importance to this area, China CDM Fund would play an increasingly significant role in pressing ahead with action and cooperation on tackling climate change and promoting energy conserving and emission reducing undertakings.

ADB President Haruhiko Kuroda remarked that China CDM Fund was a pioneering move in that it utilized income from international carbon market to eliminate domestic barriers and develop low-carbon economy. ADB would continue its cooperation with the fund in an all-round manner to promote China's work on addressing climate change.

President Haruhiko Kuroda of Asian Development Bank addressing at the opening ceremony

The launch of China CDM Fund has attracted wide attention inside and outside China.

7.1.4 The Release of *Measures for Management of China Clean Development Mechanism Fund*

Passed by administrative meetings of Ministry of Finance, National Development and Reform Commission, Ministry of Foreign Affairs, Ministry of Science and Technology, Ministry of Environmental Protection, Ministry of Agriculture and China Meteorological Administration, and approved by the State Council, *Measures for Management of China Clean Development Mechanism Fund* (the Measures) was promulgated by decree of the seven ministries on 14th September 2010. The Measures sketched an overall business blueprint for the fund including the administrative organizations and its responsibilities, fund raising, fund utilization, grant project management and investment project management, laying a policy base for the fund's operation in full swing.

7.2 The Governance Structure of China CDM Fund

7.2.1 The Purpose, Nature and Strategic Position

According to *Measures for Management of China Clean Development Mechanism Fund*, the fund is a public institution that integrates government and market functions. The purpose of the fund is to support China's efforts in tackling climate change and promote sustainable development. The fund will implement under the guidance of national sustainable strategy the tasks specified in *China National Climate Change Programme* and *Comprehensive Working Program on Energy Conservation and Emission Reduction*.

As an innovative climate financing mechanism, China CDM Fund will combine government fund, international assistance and social funding and serve as a platform for financing, action, cooperation and information gathering and exchange. The fund supports activities like capacity building, publicity, energy efficiency programs and new energy development by means of offering grants, equity investment, concessional loans, portfolio guarantee and etc. and supplements the main source of fund to support national climate work. Meanwhile with the support from the government, the fund will adopt market-oriented operation model and cooperate closely with companies and commercial financial institutions to disseminate the successful experience of pilot projects and play its due role as a public fund.

7.2.2 The Governance Structure

China CDM Fund is governed by the Board of China CDM Fund (the Board) and China CDM Fund Management Center (the Management Center). The Board is an inter-ministerial deliberative organ on fund affairs.

The Management Center, as a daily administrative organ, is responsible for collection, management and use of the funds, and operates the fund under the guidance of the Board.

The Strategic Development committee, the Risk Control Committee and the Investment Review Committee have also been established to ensure steady and sustained development of the management center, improve its governance structure, make decision-making more scientific, keep risks under control and enhance the competitiveness. The Board and the Strategic Development Committee check and balance with the Management Center on outward front while the Risk Control Committee and the Investment Review Committee balance inside the Management Center, so that the safe and effective operation of the fund can be secured in the process of project initiation, investigation, approval, risk control and supervision.

The governance structure of China CDM Fund is shown in diagram 7-1.

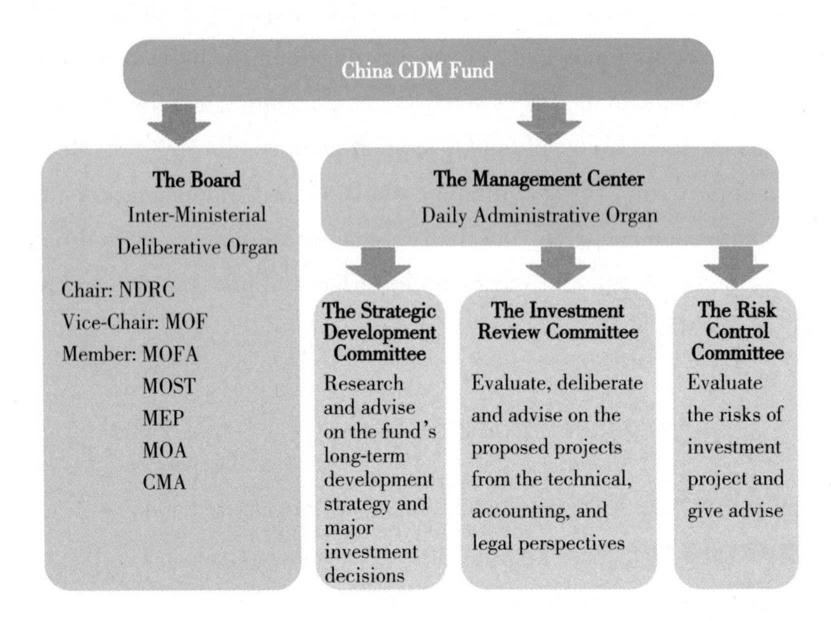

Diagram 7–1 The Governance Structure of China CDM Fund

7.2.2.1 The Board of China CDM Fund

The Board of China CDM Fund is an inter-ministerial deliberative organ, which consists of representatives from NDRC, MOF, MOFA, MOST, MEP, MOA and CMA. NDRC and MOF assume respectively Chair and Vice-Chair of the Board, and their representatives will perform the duties.

The Board reviews the following items:

(1) Basic regulation of the fund.

(2) Strategic planning of the fund, including annual budget plan.

(3) Application for grants and investments.

(4) Annual budget plan and financial accounts.

(5) Other major undertakings.

7.2.2.2 China CDM Fund Management Center

The Management Center is the daily administrative organ of the fund, responsible for collection, management and use of the fund. The Center is subject to the supervision of MOF.

The Management Center mainly performs the following duties:

(1) Draft and formulate regulations on management and operation of the fund.

(2) Raise funds.

(3) Manage the funds, and carry out investment and wealth management activities.

(4) Formulate and execute the annual budgeting and accounting plan of the fund.

(5) Supervise and manage projects sponsored by the fund.

(6) Report major business activities of the fund to the Board.

(7) Engage in other activities that serve the purpose of the fund.

7.2.2.3 The Strategic Development Committee of the Management Center

To ensure steady and sustained development of the fund, improve its governance structure, make decision-making more scientific, and enhance competitiveness, the Management Center has established the Strategic Development Committee, which will research on the mid-and-long term development strategy and major investment decisions and provide due recommendations.

The main responsibilities of the committee are as follows:

(1) Research and advise on the mid-and-long term development strategy of the Management Center.

(2) Research and advise on the major investment and financing plans of the Management Center.

(3) Research and advise on the major capital and asset operation projects of the Management Center.

(4) Research and advise on other major items that affect the development of the Management Center.

(5) Review the implementation of above-listed items.

(6) Perform other duties mandated by the Management Center.

On 16th April 2009, the first Strategic Development Committee was formed and the first plenary meeting of the committee was held. The meeting elected He Bangjing, member of CPC Central Commission for Discipline Inspection and former head of Discipline Inspection Group of Ministry of Finance as chairman, and Vice Minister of Finance Zhu Guangyao vice chairman.

7.2.2.4 The Investment Review Committee of the Management Center

The Investment Review Committee is a deliberative and executive organ of the Management Center. It is mainly responsible for investment planning, strategy proposition and evaluation and deliberation of the projects before making investment. The committee is under the supervision of the Management Center and its assessment result will be taken as a reference by the Management Center while making investment decisions.

The main responsibilities of the committee include:

(1) Review the investment management strategy and evaluation criteria, regulation and policy of the Management Center.

(2) Review the investment area, strategy and restrictions of the Management Center.

(3) Review the annual budget plan for investment of the Management Center.

(4) Conduct independently technical and economic evaluation of the investment based on the budget plan, feasibility study report, project implementation plan and due diligence report.

(5) Monitor the portfolio companies and supervise the investment activities of the Management Center.

(6) Review the disposal plan for portfolio companies of the Management Center.

(7) Review the provision plan for the risk-ridden portfolio companies of the Management Center.

(8) Research and advise on other important investment items that affect the development of the fund.

(9) Conduct on-site survey of the investment project if necessary.

(10) Perform other duties mandated by the Management Center, *Measures for Management of China Clean Development Mechanism Fund and Measures for Management of Investment Projects of China Clean Development Mechanism Fund.*

7.2.2.5 The Risk Control Committee of the Management Center

The Risk Control Committee is an ad hoc deliberative organ of the Management Center, responsible for monitoring and managing the fund's systematic and structural risks as well as the investment risk.

The main responsibilities of the committee include:

(1) Review the risk management strategy and systematic planning of the fund.

(2) Evaluate the investment risks and provide risk-tackling recommendations. The risk evaluation report of the committee will be deemed as a necessary reference for assessing investment project. The report will be submitted to the Investment Review Committee for reference and the Director General's Meeting for deliberation and approval.

(3) Guide the work on improving the systematic and institutional design for better risk management structure, procedures, and risk disposal method.

(4) Perform other duties mandated by the Management Center, *Measures for Management of China Clean Development Mechanism Fund* and *Measures for Management of Investment Projects of China Clean Development Mechanism Fund.*

In addition, to ensure scientific and rational development, evaluation, implementation and supervision of the investment projects, the Management Center has established a Think Tank, consisting of six types of experts from policy, industrial, financial, risk management and legal areas. To ensure that the experts' advice is fair and just, the Management Center selects randomly experts from the Think Tank to assist the Investment Review Committee and the Risk Control Committee in case of need.

Chapter Ⅷ

The Main Undertakings of China CDM Fund Management Center

In accordance with the State Council mandate and *Measures for Management of China Clean Development Mechanism Fund,* the Management Center has identified three main undertakings: fund raising, capital management and fund utilization.

8.1 Fund Raising for China CDM Fund

8.1.1 Sources of Fund

Raising fund from various sources to expand fund scale is key to the fund's operation. Pursuant to *Measures for Management of China Clean Development Mechanism Fund,* the fund will be raised mainly from the following four sources:

(1) National revenue from the CERs transfer benefit generated by CDM projects.

(2) Operating revenue of the fund.

(3) Donations from institutions, organizations and individuals from home and abroad.

(4) Other sources, including financing in the market, fiscal allocations and etc.

At present, the main source of the fund is national revenue, thus the next part will focus on national revenue and its collection.

8.1.2 National Revenue

The Management Center has opened a dedicated account to collect national revenue. National revenue is collected and managed in ways similar to fiscal revenue and fully incorporated into the fund.

8.1.2.1 Setting of National Revenue

The CDM project owners shall pay national revenue fully and timely in accordance with *Measures for Management of China Clean Development Mechanism Fund* and *Measures for Operation and Management of Clean Development Mechanism Projects* after they obtain revenue from transfer of each batch of CERs issued by the CDM Executive Board (EB). The Management Center will adhere to the principle of equity, openness, fairness and operability in collecting national revenue.

National revenue will be calculated in the following formula:

$$\text{National revenue} = \left(\text{CERs issued by the EB} - \text{donated CERs to the Adaptation Fund}\right) \times \text{unit price} \times \text{national revenue levy rate}$$

Donated CERs to the Adaptation Fund amount to 2% of CERs issued for a CDM project activity as required by the UN CDM rules.

Unit price refers to CER transfer price agreed in the emission reduction purchase agreement (ERPA) contracted between the project owner and the buyer submitted to DNA for application.

National revenue levy rate is specified in *Measures for Operation and Management of Clean Development Mechanism Projects (Revised Edition)* as follows, taking effect from 3rd August 2011:

(1) The Government of China takes 65% CER transfer benefit from HFC projects.

(2) The Government of China takes 30% CER transfer benefit from N_2O projects related to adipic acid production.

(3) The Government of China takes 10% CER transfer benefit from N_2O projects

related to the production of nitric acid and etc.

(4) The Government of China takes 5% of CER transfer benefit from PFC projects.

(5) The Government of China takes 2% of CER transfer benefit from other CDM projects.

8.1.2.2 Collection of National Revenue

A. Collection Process

(1) Submitting Transaction Information for Filling. The project owner submits to the Management Center, a copy of ERPA, and a copy of Business License, contact information of both contracted sides within 15 working days after the effective date of the ERPA. In case of any change to the filed information, the project owner shall notify the Management Center in written form within 15 working days after the date of change and submit relevant certifying documents.

(2) Submitting Payment Confirmation Letter by the Project Owner. The project owner submits in written form to the Management Center the Payment Confirmation Letter of National Revenue of CDM Projects (the Payment Confirmation Letter) with company seal within 15 working days after the date of CER issuance and transfer (The Payment Confirmation Letter is made by the Management Center and downloadable from its official website).

(3) Paying National Revenue by the Project Owner. The project owner pays national revenue to the dedicated account within 15 working days after obtaining benefit from each CERs transfer. In case of ERPA specifying otherwise, the buyer of CERs shall make direct payment of national revenue to the special account for national revenue.

(4) Issuing receipt by the Management Center. The Management Center shall issue receipt to the payer within 15 working days after receiving national revenue in full amount and certifying documents. In case of CER buyer paying national revenue as agreed in ERPA, the Management Center shall issue documents certifying the paying of national revenue to the project owner after receiving national revenue in full amount.

The project owner shall note the following issues in the process:

(1) National revenue could be paid in the currency stipulated in ERPA or in RMB. If the payment is made in RMB, the exchange shall be calculated based on the buying rate at the time of settlement, and a copy of settlement voucher with the company's finance-only seal should be submitted together.

(2) If the buyer makes the payment with tangible assets such as equipment or intangible assets like patent and technology, the project owner calculates national revenue according to price contracted in ERPA, and make the payment in RMB within 15 working days after the date of CER issuance and transfer.

(3) If the price contracted in ERPA is not fixed, the project owner submits a copy of invoice about the CERs transfor with company seal.

B. Penalty

In accordance with Article 31 of *Measures for Operation and Management of Clean Development Mechanism Projects (Revised Edition)*, if the project owner does not make full or timely payment of their share of national revenue from CDM projects after the trading of CER, NDRC will impose administrative penalties.

8.1.2.3 Estimated Scale of National Revenue

In view of the current development of CDM projects in China, if the international and domestic policies remain stable and China's CDM projects remain in smooth operation before 2012, it is estimated that by the end of 2012 the accumulated project revenue for China will reach USD 8.5-10 billion, among which national revenue will reach USD 1.9 billion (RMB 12.2 billion[1]). HFC-23 projects will contribute 80%, N_2O projects 12% and other types of projects 8% to national revenue.

The post-2012 national revenue will depend on the negotiation result over the second commitment period of Kyoto Protocol, the fact whether the UN CDM rules will change, the policies of international buyers like EU and Japan as well as China's climate policy.

1 Here USD to RMB exchange rate is assumed to be 1:6.4.

8.2 Utilization of China CDM Fund

By the nature of the fund's capital, there are mainly two ways to utilize the fund. The first is offering grants to sponsor the urgently needed climate activities like policy research, capacity building and publicity. The second is to fund industrial activities conducive to combating climate change by making investments.

8.2.1 Grant

The fund offers grants to mainly support following activities:

(1) Policy research and academic activities on climate change;

(2) International cooperation on climate change;

(3) Trainings to strengthen capacity building for tackling climate change;

(4) Publicity and educational campaigns to enhance public awareness about climate change;

(5) Other activities that serve the purpose of the fund.

Grants shall not be used to support profit-seeking activities or cover government's and public institution's expenditure, and in principle grants will not sponsor activities that have already received capital support from other sources.

Management and use of the grant include the following:

8.2.1.1 Organize Project Application

Based on the need of climate work, the Board of China CDM Fund will allocate part of fund out of the annual budget as grants of the year. And it will also identify key grant-supporting areas and organize the application of grant projects.

8.2.1.2 Project Application

The grant project applicant shall prepare in view of the key areas and requirements specified by the Board an application proposal for grant project, which must include following items: the profile of the applicant, project background, project goal, project activities, project output, timetable for implementation, applied grant amount, budget plan and other items concerned. This statement shall be transferred or submitted to the NDRC by relevant department of the State Council or provincial Development and Reform Commission.

The grant project applicant shall be the relevant domestic institution of climate area capable of undertaking research or training.

8.2.1.3 Project Review

The NDRC after receiving the application proposal will organize a group comprising financial, managerial and technical experts to review the qualifications of the applicant, completeness of the proposal necessity and feasibility of the project, justifiability of the claimed grant amount and other items, and then the group will formulate and submit the review opinion to the Board of China CDM Fund for deliberation and verification. On the basis of the review opinion of the expert group, the Board will verify the project and put forward its opinion of shorted-list projects and their corresponding grants. The NDRC and MOF will jointly approve based on the opinion of the Board the annual list of grant projects and the amount of grants.

8.2.1.4 Contract Signing

After the grant project receives joint approval from the NDRC and MOF, the NDRC, the organizer of project application, the Management Center and the project applicant shall enter a joint contract for grant project that specifies the rights, obligations, liabilities and penalties for breach of contract.

8.2.1.5 Project Implementation, Supervision and Management

The applicant of grant project shall implement the project in strict accordance with the contract, and shall be subject to guidance from the NDRC, the Management Center and the organizer of project application. The NDRC, the Management Center and the organizer of the project application will be responsible for supervising and inspecting the project implementation. Irregularities in the implementation process will be penalized and punished by NDRC and MOF pursuant to relevant regulations.

8.2.2 Investment

According to the provisions of *Measures for Management of China Clean Development Mechanism Fund*, the fund will sponsor industrial activities conducive to combating climate change via investment. The investment will be focused on industrial activities that can not only help significantly reduce emission of greenhouse gases, but also can yield good economic returns and have sound demonstrative effect. Meanwhile, the fund will fully play its guiding and catalyzing role to actively explore innovative financing mechanism that can promote industrialization, marketization and socialization of China's work on tackling climate change, conserving energy and reducing emissions, and channel various resources to the climate undertaking, to support the country realize low carbon goals.

8.2.2.1 Ways of Investment

Ways of investment mainly include equity investment, concessional lending, portfolio guarantee and other ways ratified by the state.

According to the fund management measures, the annual investment in projects via obtaining equity or providing concessional loans shall not exceed a certain percentage of the fund's net assets of the previous year. The underlying assets of guarantee shall not exceed the limit set by the fund's annual budget plan. The fund shall not seek controlling status over the investee when making equity investment and when the investment exits, the way and price of the exit shall be decided in open, fair and market-based principle.

The fund shall not make investment in stocks, stock-investing mutual funds, real estate, futures and other financial derivatives.

8.2.2.2 Project Application

The investment project applicant must be a Chinese enterprise or Chinese equity controlled enterprise that engages in business of mitigation and adaptation to climate change or of relevance. The applicant shall file an application with the Management Center, and submit the following documents:

(1) The application proposal;

(2) The feasibility study report of the project;

(3) The report of the company's performance over the last three years;

(4) The business license of the company;

(5) Other documents concerned.

8.2.2.3 Project Operation and Management

The Management Center will take the national economic, industrial and climate policies into consideration and select and review the investment projects within the investment and guarantee limit set by the budget plan. According to the management measures of the fund, each project with applied capital over RMB 70 million is classified as major project, which shall be submitted by the Management Center to the Board for opinion and must obtain approval from the NDRC and MOF based on the consensus of the Board; each project with applied capital below RMB 70 million is classified as ordinary project, which shall be approval by the Management Center and submitted to the NDRC and MOF for filing.

The Management Center will be responsible for organization of project implementation, project supervision and inspection, and examination and acceptance of the project. Assets, rights and interests produced by investment will be managed in conformity with relevant financial regulations and rules of China.

8.3 Capital Management of China CDM Fund

Capital management involves revenue collection, fund utilization and capital expenditure, namely management of various revenues, capital expenditure for grant projects, fund utilization for investment projects, and capital flow of wealth management activities.

8.3.1 Management of Fund Revenue

The revenues will be fully collected and fully deposited in the bank account designated by the fund.

8.3.2 Capital Management for Grant Projects

Capital of grant projects refers to the fund offered as grants to projects after being verified by the Board and jointly approved by the NDRC and the MOF.

The institution implementing the grant project must compile budget plan with detailed description of the purpose of fund using in accordance with the project's capital regulations, and submit the plan to the Management Center. Managing grant projects on the basis of contract, the center will deliver the grants in ways as agreed in the contract.

In general, grant appropriation will be made in three phases based on project progress.

(1) After signing contract, the Management Center will transfer 40% of total grants approved and specified in budget plan, as reserve fund in the contract.

(2) After the project implementer submits the mid-term reports (fund utilization report included), and NDRC, the project application organizer and the Management Center verify and consent to the reports, the Management Center will transfer another 40% of total grants approved and specified in budget plan.

(3) After NDRC, the project application organizer, the Management Center confirm the acceptance of the project and financial auditing compliance, the Management Center will transfer the rest of grants within 15 working days to the project implementer.

Grants shall be included in the overall financial plan of the project but meanwhile accounted separately. In case of government procurement, accounting shall be done according to the regulations governing government procurement. If there are legal items stipulate otherwise, the upper legal items shall prevail.

In the process of project implementation, the Management Center will monitor the financial status of the grant project, and update the Board timely. Besides, the Management center will organize the project implementer to carry out financial self-examination and also arrange for qualified accounting firms to audit the project implementer and the project's mid-term accounts and final financial results; so that it can improve its capital management of grant projects to realize the purpose of offering grants.

8.3.3 Capital Management for Investment Porjects

Capital of investment projects refers to the fund earmarked by the Management Center for sponsoring industrial activities conducive to combating climate change as is mandated by *Measures for Management of China Clean Development Mechanism Fund.*

The Management Center will make investments within the limit of the fund's annual budget plan approved by the MOF and NDRC. And it will make timely and accurate statements about the financial status, results of capital utilization, formation and disposal of and return on assets of investment projects pursuant to the general accounting principles. The project implementer shall use the capital in light of the content, purpose, timetable and progress of the project set in the contract. After the project ends, the Management Center will assess the project's capital utilization, return of capital and project performance, and submit the assessment results to relevant administrative authorities and the Board of China CDM Fund.

8.3.4 Cash Management Activities

In order to play the role of the fund to the full, and maintain and increase the fund value, the Management Center will carry out cash management activities with its surplus funds.

The Management Center shall follow principles of safety, liquidity and growth, and the purpose of promoting energy conservation, energy efficiency improvement, new energy and renewable energy in conducting wealth management activities, including making structural deposits, purchasing treasury, financial, and high-rating corporate bonds, as well as engaging in other low-risk financial activities. Wealth management activities will be carried out solely in cooperation with domestic commercial banks for controlling risks.

By carrying out cash management activities, the fund can not only play its role as seed capital to attract more social funds, but also activate other domestic capital for low-carbon development and improve capital utilization efficiency, and more importantly, it can standardize statistics of emission reduction and help enhance the companies' capability of reducing emission.

Chapter IX

The Completed and On-Going Work of China CDM Fund

Since its inception the Management Center has placed due importance on fund raising and fund management and utilization, and has been well positioned to realize its purpose of "building itself into four platforms (platforms of capital, action, information and communication), and serving the low-carbon development". The center has carried out in an all-round manner its undertakings in the areas of formulation of charter and regulations, collection of national revenue, management of grant projects, development and management of investment projects, research on major issues, wealth management of the fund's capital, serving corporate low-carbon development, strengthening international cooperation and enhancing public awareness and etc, which has served China's work on tackling climate change, conserving energy and reducing emission well.

9.1 Establish Charter and Regulations to Ensure Standardized Operation of the Fund

Since the beginning of preparations for the fund, the formulation of charter and regulations and capacity building have been on the priority list of the Management Center, in order to ensure scientific, standardized, and high-criteria operation of the fund. Therefore, the Management Center has established a series of charter, regulations, rules and norms governing organizational, administrative and financial management, and human resource and project management of the fund. And this has been an effective guarantee to safe and orderly operation of a newly-established institution.

On 14th September 2010, *Measures for Management of China Clean Development Mechanism Fund* was officially unveiled and it has clearly defied the nature, purpose, principle of utilization, managing institution of the fund, and specified rules governing fund raising and using. Based on the Measures and under the guidance of the Board of China CDM Fund, the Management Center has formulated a series of more detailed regulations, including *Rules on Meeting of the Board of China Clean Development Mechanism Fund, Measures for Management of Grant Projects of China Clean Development Mechanism Fund, Measures for Management of Investment Projects of China Clean Development Mechanism Fund, Measures for Financial Management of China Clean Development Mechanism Fund, Measures for Collection of National Revenue of China Clean Development Mechanism Projects, Measures for Management of Concessional Loans of China Clean Development Mechanism Fund, Measures for Management of Equity Investment of China Clean Development Mechanism Fund, Measures for Management of Portfolio Guarantee of China Clean Development Mechanism Fund, etc.*. At present, regulations and norms governing the business and organizational operation of the fund and the Management Center have been generally well in place.

9.2 Collect National Revenue to Underpin the Fund's Undertakings

For the moment, national revenue from transfer of CERs generated by CDM projects is the main source of revenue for the fund. Therefore, to collect national revenue in full amount and in time is fundamental to the fund's development. The Management Center has made remarkable achievements in this regard.

9.2.1 Standardize the Working Process to Secure National Revenue

As the collection of national revenue involves multiple links, long process, diverse target groups of levy, and heavy tasks, the Management Center has formulated *Implementing Rules for Collection of National Revenue of China Clean Development Mechanism Projects* to guide the work with a detailed and standard internal and external process, designed and developed the management information system on national revenue collection, and all these efforts have ensured a sound collection work.

9.2.2 The Collected National Revenue

From 13rd April 2007, Taonan Wind Power Project of Jilin Province paid in the first national revenue of RMB 96.47 thousand, until 30th June 2011, the Management Center has received RMB 8.042 billion as national revenue from transfer of 736 batches of CERs generated by 370 projects.

Among the above-mentioned national revenue, the HFC-23 projects have contributed RMB 7.253 billion, accounting for 90.2% of the total. That's because the HFC-23 projects are early in implementation, large in emission reduction, and high in levy rate, that is 65% of CERs transfer benefit. So far all this type of projects have been developed. As other types of projects unfold gradually, the contribution of HFC-23 projects to national revenue is going to decline. Similarly, N_2O projects, another industrial gas project, though started relatively late, yet produce comparatively massive emission reduction and with a levy rate of 30%[1], this type of project has become a main source of national revenue, having contributed RMB 694 million representing 8.6% of total national revenue. New energy and renewable energy projects like wind power, hydropower and bio-mass projects, and energy conservation and efficiency projects accounting for 95.3% of all CDM projects with

1 *Measures for Operation and Management of Clean Development Mechanism Projects (Revised Edition)* issued on 3 August 2011 adjusted the ratio of national revenue for N_2O decomposition CDM projects: for adipic acid production is 30%; for nitric acid production is 10%.

issued CERs have generated 83.5% of all batches of CERs. But as those projects are small in scale and are encouraged by the state with a low levy rate of 2%, national revenue contributed by them are merely RMB 96 million, posting 1.2% of total national revenue (show in Table 9–1) .

Table 9–1 National Revenue Collected from Different CDM Projects

Sector	Levy Rate (%)	Number of Projects	Number of Issuance	Accounts Received (RMB million)	Percentage (%)
HFC decomposition	65	11	104	7,253.086	90.2
N_2O decomposition	30	7	38	693.5384	8.6
New energy and renewable energy	2	245	460	55.6504	0.7
Fuel switch	2	12	26	19.80.93	0.2
Methane recycling	2	13	31	13.5692	0.2
Energy conservation and energy efficiency	2	33	65	6.7168	0.1
Aggregate		489	1,125	8,042.3701	100.0

Note: updated on 30th June 2011.

Source: Data collected by China CDM Fund Management Center.

The proportions of national revenue shows that industrial gas projects have contributed a bulk of national revenue, which testifies that the government of China is far-sighted and responsible on the issues of climate change, in that Chinese government set up China CDM Fund to regulate the CDM revenue and make the company's benefit from industrial gas projects "green". This move alleviates international concerns for potential ill consequences of the HFC-23 and N_2O projects, and enables the Clean Development Mechanism to play a higher-level, more comprehensive and more enduring role in China's efforts of combating climate change and promoting sustainable development.

9.3 Conduct Cash Management Business to Ensure the Safety and Value Preservation and Increment of the Fund

Since its inception, the Management Center has on the one hand adopted strict management over the fund's capital for safety purposes, and on the other hand actively taken various wealth management measures to maintain and increase the fund value as is required by the State Council and for the sustainable development of the fund. Those measures include making banking deposits of RMB and foreign currencies, purchasing treasury, financial, and high-rating corporate bonds, as well as engaging in other low-risk financial business. By Oct.2011, the center has made RMB 400 million in profits via wealth management activities.

9.3.1 Foreign Currency Denominated Wealth Management Activities

The US dollar(USD), euro and other foreign currencies account for more than 50% of the national revenue. Euro depreciated drastically due to the recent financial crisis and RMB appreciated continuously against USD. In response, the Management Center, with close watch on the economic developments of major economies and the trend of foreign exchange market, tends to settle its foreign exchanges both in real time and at selected points based on its best calculation to avoid exchange losses. According to initial statistics, the Management Center avoided exchange losses worth hundreds millions RMB during the period of continuous appreciation of RMB. Meanwhile, those follow-up efforts also can serve as a sound base for wealth management activities as the center can ride on the market trend to retain certain foreign currency for cash management in its favor. Through working with Chinese banks, the center has gained RMB several million of profits from foreign currency denominated wealth management activities like buying short-term euro products, making currency swap transactions and etc.

9.3.2 RMB Denominated Wealth Management Activities

Given that the fund has massive RMB capital and sufficient surplus capital, the Management Center has been actively engaging in risk-free wealth management activities like buying financial products at regular intervals and making call deposits at banks. The center used the surplus capital to carry out mid-and-long term wealth management activities as much as possible, which can avoid leaving idle capital unused and bring hundreds of millions of cash returns.

Apart from seeking financial returns, the Management Center also endeavored to enhance social benefits of the fund. After consultations with many domestic commercial banks, the center has proposed a contract-based cooperation model featuring "framework agreement + implementation agreement", adopted a "1 : N" leverage, and served as seed capital to divert more social funds to activities of energy conservation and emission reduction. In 2010 and 2011 the Management Center conducted cooperation with China Zheshang Bank and Industrial Bank respectively in the area of wealth management for clean development, which not only kept the fund capital safe but also helped attract 100% more of commercial capital to support china's climate efforts. It is estimated that those wealth management activities can lead to reduction of CO_2 by over 600,000 tons, and mark the first batch of capital dedicated for wealth management, and a bold and innovative endeavor of the center in combining wealth management and efforts for tackling climate change, conserving energy and reducing emission. The significance is: using innovative mechanism and playing the role of market to mobilize social funds to support undertakings of tackling climate change, conserving energy and reducing emission; optimizing advantage mix to reduce risks of the fund's capital operation and cultivate low-carbon awareness and capability of financial institutions; actively exploring a scientific and professional capital management model; combining capital and technology to promote low-carbon development of industries and support the reform and innovation of development model.

9.4 Offer Grants to Support National Efforts in Tackling Climate Change

Climate work has been incorporated in the government's agenda of important issues. However, China still has very weak work, capability and awareness bases that are imperative to be strengthened. Particularly when climate change has become an issue of great importance and a hot topic for international meetings, China needs urgently to enhance its climate capacity and awareness to meet the needs of its climate work at home and abroad. Against this backdrop, the fund can bring its unique advantage into full play by choosing and finance the urgently-needed activities via grants.

9.4.1 Grant Projects of the Fund

In 2008, the Board of China CDM Fund approved the first batch of 15 grant projects (Table 9–2), involving a total of RMB 26.55 million of budget. In 2011, the fund has allocated about more than RMB 200 million to support China's climate efforts and low carbon economy, which include over 100 projects such as the formulation of climate plans and the complication of GHG inventories of 31 provinces, municipalities and cities, the planning of low-carbon pilot programmes in five provinces and eight municipalities and cities (provinces of Guangdong, Liaoning, Hubei, Shaanxi and Yunnan, and municipalities and cities of Tianjin, Chongqing, Shenzhen, Xiamen, Hangzhou, Nanchang, Guiyang, and Baoding), formulation of plans for carbon trade pilots in five provinces (cities of Beijing, Tianjin, Shanghai, Chongqing, Shenzhen and provinces of Hubei and Guangdong) , and research on issues related to climate change by relevant ministries.

Table 9–2 List of grant projects in 2008

No.	Project Name	Finance from the Fund (RMB million)
1	Review CDM projects	6.5
2	Update *China National Climate Change Programme*	2.0
3	Research on national climate change strategy	3.0
4	Research on the interest demand, negotiating position and policies of major signatories	1.5
5	Methodology development for scaling up the large and shutting down the small in building power plants	1.0
6	Strategic research on controlling HFC-23 emission in China	0.5
7	Make a campaigning film about climate change	0.5
8	Research on and assessment of emission reduction potential in agricultural sector and in China's rural area	3.38
9	TV series: China Action on Climate Change	0.5
10	Consulting Activities of National Panel on Climate Change	0.5
11	Research on the post-2012 climate financing system design	0.6
12	Fiscal policy research on climate change	0.6
13	Capacity building for China CDM Fund	0.6
14	Institutional building for China CDM Fund	0.6
15	Research on climate scenarios, implications and policy response under different long-term stable GHG concentration goals	4.75
Aggregate		26.53

9.4.2 Accomplishments of Grant Projects

The allocated grants have supported four areas related to China's climate efforts, including international negotiation, policy and decision making, publicity and education, and theoretical research, so as to provide the much-needed scientific, theoretical, and public opinion support in a multi-faceted, multi-layered and comprehensive manner. These the grant projects have made great impacts on the following aspects:

First, some grant projects have provided important scientific base for the state to formulate climate policies and incorporate climate work into the overall development planning of the nation. For instance, the projects sponsored the updating of *China National Climate Change Programme*, and both have become very important guiding document on climate work. Some other grant projects, for example, research on climate scenarios, implications and policy response under different long-term stable concentration goals, and research on and assessment of emission reduction potential in agricultural sector and in China's rural area, have also produced important basic research results to serve the nation's decision making.

Second, some grant projects, for instance the sponsorship to consulting activities of National Panel on Climate Change, research on the interest demand, negotiating position and policies of major signatories, and research on national climate change strategy, have produced gratifying results and provided important theoretical and technical support to enable China's active, positive and effective participation in international climate talks and safeguard the interest of China and other developing countries.

Third, some projects' activities and results, for instance the campaigning film about climate change and the TV series of China Action on Climate Change, have helped to make China's climate efforts visible domestically and internationally, so that the international community better understands China and the public better aware the issue of climate change. And this has brought very positive international and social impact.

Fourth, some projects, for instance the review of CDM projects, have lent very strong support to the development of CDM projects in China, so that China has witnessed the quick spring-up of CDM projects across the nation. Soon China ranked first in the world in terms of three major indicators, the number of projects approved by the sate, projects registered at EB, and CERs issued, and became one of the key carbon markets in the world.

Being well-targeted, profoundly meaningful, and quick in approval, those grant projects have provided timely and strong support to the policy and strategy formulation of Chinese government, to the participation and organization of domestic and international climate meetings in 2008, 2009 and 2010, and to the climate publicity inside and outside China. This testifies to the unique and important role of the fund.

9.5 Actively Promote Fund Investments to Support Energy Conservation and Emission Reduction

The State Council has clearly defined the fund as "a public fund that integrates government and market functions", which is also a basic positioning for the fund. The fundamental objective and task of the fund is to promote scientific, market-based and professional way to conserve energy, reduce emission and combat climate change by innovative mechanism. As a public fund, it will reflect government policy direction. As a fund that integrates market functions, it can operate in the broad market and engage in flexible cooperation with various commercial institutions. The combination of two bridges the gap between the government policy and market development. The Management Center has been exploring since its establishment the strategy, key areas, ways of investing for the fund's investment projects, and have been well prepared to support national policies of conserving energy and reducing emission through investment.

9.5.1 Clearly Define the Strategy for Fund Investment

At present, the fund has a total of RMB 8 billion under management, and it is expected to reach RMB 12.2 billion by the end of 2012. Yet compared with fiscal input in climate work, energy conservation and emission reduction efforts, and the capital inflow into the relevant areas from financial institutions and businesses, the fund capital still amounts to minimum and is utterly inadequate However, as an innovative financing mechanism, the fund shall take advantage of new mechanism, new organization and new members, to play an innovative, pioneering, guiding, and demonstrative role in China's work of developing low-carbon economy and combating climate change.

After visiting lots of financial institutions, commercial banks, local governments, and enterprises, and also taking reference from the models of World Bank, Asian Development Bank, Social Security Fund, China-Belgium Fund and Carbon Trust, the Management Center identified its long-term objective and task as to support the industrialization, marketization and socialization of China's work on tackling climate. The center solved the problems of finding impetus and method for the fund's development and attracting wide participation from various social groups. It also proposed the strategic guiding principle of "sailing in borrowed boat, operating by professionals, and developing with partners". Namely the fund will use its limited fund as seed capital to divert more social resources to energy conservation and emission reduction and earnestly pursue scientific, innovative and sophisticated development. Meanwhile, partnership with social forces can solve the problems of weak capacity of controlling market and moral risks in the part of the center, which is a supplement to the fund's weakness.

Given the reality that most emission comes from the industrial sector and industrial production is low-tech, backward in process and management and thus huge potential for reduction, the Management Center proposes to make investment with focus on three types of reduction, i.e."market-based, technical and social reduction". The three types of reduction is an embodiment of the current work to achieve fund's long-term objective of "industrialization, marketization and socialization of China's work on tackling climate change, conserving energy and reducing emissions", and also an action guideline.

As a public fund, instead of a commercial financial institution, the fund does not seek maximum economic interest and the Management Center values particularly the social return of the fund investment. When the center reviews the investment projects, it will consider, apart from economic return and risk, emission reduction as one of the important indicators, and require the applicant to submit an estimated carbon reduction report along with other application documents. In the process of implementation, the projects must keep books of carbon assets and calculate the reduction. This not only supports the nation's work on energy conservation and emission reduction, but also enhances the recipients' awareness of managing carbon assets, and this reflects the guiding role of the fund.

9.5.2 Explore Ways of Fund Investment

Regarding concessional lending, the Management Center attempts to provide, in cooperation with local governments, commercial banks and large-scale state-owned enterprises, the eligible companies with consignment loans at a rate lower than the corresponding benchmark interest rate of the People's Bank of China (PBC) for a certain period of time. At present, the fund has first launched the concessional lending model in cooperation with local governments. By 21st October 2011, concessional loans of over RMB 1 billion have been extended to projects in eight provinces including Shaanxi, Shanxi, Hebei, Hunan, Fujian, Jiangsu, Shangdong and Jiangxi, and after coming into production, the projects are estimated to reduce emission of greenhouse gases by 10 million tons of carbon dioxide equivalent (CO_2e) annually.

Regarding equity investment, the Management Center will comply with the national industrial policies, take both the capital market condition and the status of the fund into consideration, and utilize financial institutions' customer base and rich experience to select high growth enterprises for its equity investment.

Regarding financing guarantee, the Management Center is stepping up its exploration of business model for financing guarantee together with professional guarantee agencies, and is also studying and discussing cooperation with them, so that apart from investment, the fund's capability of operating projects will be enhanced and improved as well.

Case Study: Concessional Loans for Clean Development Wind Farm Project of Company C in City X of Fujian Province

Company A of Fujian Province is a large-scale state owned enterprise with a registered capital of RMB 5.9 billion. By the end of 2010, the total asset of the company was RMB 38.9 billion, and among that the owner's equity was RMB 22.4 billion. In December 2008, Company A (51% of share) and Company B of Fujian Province (49% of share) invested RMB 80 million together to set up Company C, which is investment firm on wind power projects.

In order to develop and utilize the rich wind energy resources along the coastal region, ease the reliance on primary energy, alleviate environmental pollution, and also serve as a pilot program to guide Company A's future massive development of wind power projects, Company C plans to invest RMB 600 million to build a wind power project in City X. the project plans to install twenty-four 2MW permanent magnet direct drive wind turbines from the Netherlands's Vestas to reach on-grid generating capacity of 48MW, and meanwhile a 110kV low-high substation. After conducting on-site wind tests, the project started construction in February 2011, and is expected to start production by the year end. To partly relieve financial pressure, the Fujian Investment & Development Group has asked the Fujian Provincial Department of Finance to apply for RMB 70 million of concessional loans from the fund, accounting for 12% of total investment.

(1) Social Return. New energy development including the utilization of wind power is in conformity with the national industrial policies and the requirement of sustainable development. Full exploration of China's rich wind power is conducive to less

reliance on primary energy, rational development and optimal allocation of energy resources, and coordinated development of energy and environment. Moreover, it improves China's energy mix apart from creating more jobs.

(2) Environmental Return. This project is in compliance with national energy development trend and power development planning of Fujian Province. According to the estimation, after coming into production, the project can provide 138GW·h of electricity to the grid, marking a reduction of carbon emission of 112,300 tons. The project also reduces many air pollutants and a large quantity of dust and ashes, which helps improve air quality.

(3) Economic Return. After coming into production, the project will generate RMB 71.95 million in sales revenue and 12.1635 million in net profits.

As investment in wind farm has a long payback time and requires relatively large initial investment, Company C in order to reduce risks proposes to ask its parent company, Company A, to pay back the loans. Company A realized 732 million in net profits. With an analysis into the company's operational performance in the next three years, the company is deemed to be able to pay back the loan on time.

After going through internal review procedures of assessment by Risk Control Committee and Investment Review Committee, and submitting the project to Ministry of Finance and NDRC for filing, the Management Center finally decided to provide Company C with RMB 50 million of loans at a rate of 5.6525% for three years, and this attracted another 11 times of lending amount from social channels.

9.6 Research on Policies to Function as a Think Tank

Since its founding, the Management Center has been following closely the international cooperation on climate change, the development of international carbon markets, and domestic actions on climate change and low-carbon development. Bearing its development need in mind, the center has organized a series of policy research so as to provide advice for China's policy formulation and actions in the areas of climate change and low-carbon development, and play the innovative role of the fund.

9.6.1 Research on Climate Financing and Market-based Emission Reduction Mechanism

Financing is a key issue in tackling climate change and conducting international cooperation on climate change. The Management Center has published *Climate Change Financing*, which is the first book in China to expound on the climate financing issue, based on its accumulated policy research results in the process of preparing and operating the fund. The center also offers advice, in light of the fund's position, on promoting scientific and professional management of the capital for conserving energy and reducing emissions.

The Management Center has been actively tracking the international actions that utilize market mechanism to tackle climate change and achieve low-carbon development, including the development of global carbon market and China's carbon trade, and submitting the latest information and analyzing reports to government departments for their reference or publishing in magazines, newspapers and websites.

9.6.2 Research on MRV to Press for the Building of "Three Platforms" in Domestic Carbon Market

To reflect the guiding role of the fund, enhance added value of the fund's investment, and explore the building of infrastructure for domestic carbon trade scheme, the Management Center has organized well-targeted policy research, and cooperated with industrial entities in drafting demonstrative reduction standards for the energy intensive industries to formulate their own standards on emission reduction. The center has also made preparations for establishing a verification and certification platform and has initiated carbon budgeting, evaluation and certification for the fund's investment projects. Preparations have been made for building certification platform, transaction platform and clearance platform respectively to facilitate the establishment of a domestic carbon market. All these efforts are made to meet the needs of ensuring reduction to be "measurable, reportable and verifiable" (MRV).

9.7 International Cooperation

As a fruit of international cooperation on climate change, the Management Center gives full play to this advantage, introduces advanced concept, cut-edge technology and practical experience to China, conduct capacity building, information and knowledge sharing, so as to facilitate China's climate work.

The Management Center has attended for several times the United Nations Climate Change Conference, Carbon Expo, Carbon Asia Forum, Clean Energy Week of ADB and other activities. And by holding exhibition and sideline meetings, delivering speeches at meetings, accepting interviews with foreign journalists and other means, the center has introduced its role in China's efforts of combating climate change and pursuing low-carbon development to the international community and enhanced its recognition and influence in the global context.

9.8 Publicity for Enhancing Public Awareness of Low-Carbon Development

Tackling climate change and pursuing low-carbon development need public participation. Enhancing public awareness is of great significance, and is deemed to be an important task by the fund. Since its founding, the Management Center is actively carrying out publicity activities, like building official website, making internal publications, attending important domestic and international conferences, accepting press interviews, engaging in business exchanges, launching grant projects and etc.

The Management Center has launched its official website (http://www.cdmfund.org), which has become one of the major websites dedicated to publicity on climate work in China. The Management Center often briefs the public on the latest events of domestic and international work on climate change and low-carbon development, as well as the latest developments of the fund's work through newspaper, TV and other media.

Over the years, the Management Center has been holding workshops on climate change and low-carbon development within the government's financial system across the nation, to enhance the local financial departments' awareness of this issue and meanwhile build base for the center's cooperation with them.

Chapter X

The Prospects of China CDM Fund

At the important strategic transition period of social and economic development, Chinese government has put forward "Five Adherences" to guide the next five-year reform, requiring that the "transformation of economic development pattern" be fully intensified, scientific outlook on development upheld and harmonious and sustainable development of society promoted. Climate change has an impact on environment and resource potential for social and economic development. As a new financing mechanism to tackle climate change, the fund will comply with the national unified planning, fully play its platform role, and actively guide, support and facilitate comprehensive and systematic climate actions to promote the industrialization, marketization and socialization of China's work on tackling climate change, conserving energy and reducing emissions.

First, it is key to keep stable and safe operation of the fund. The Management Center, apart from collecting national revenue in full amount and in time, will on the one hand strive to find new source of revenue for the fund, on the other hand actively explore new financial management models to enhance efficiency, control risks and maintain and increase the value of the fund.

Second, innovation, support, guidance, and catalysis are the major themes of the fund's undertakings. Regarding grant projects, the fund will supplement the fiscal input and rationally allocate grants to support the national and local governments' policy research, capacity building and publicity activities and etc. Regarding investment projects, the fund will focus on areas of clean energy, renewable energy, industrial energy conservation and etc., sponsor projects with great reduction potential, good economic return and obvious demonstrative effect by means of investment, lending, guarantee and etc, and divert more social funds to the efforts of combating climate change and pursuing low-carbon development.

Third, as a complement to national energy efficiency auditing, the fund, with the objective of achieving market-based reduction, technical reduction, and reduction of emerging industries, will use its unique platform advantage to explore ways of policy and market mechanism innovation together with renowned research agencies at home and abroad and promote the establishment of domestic carbon trade market during the 12th Five-Year period. The fund will require each investment project to make estimated carbon reduction plan and carry out carbon stocking and via those projects, the fund can guide and enhance public awareness and capability of managing carbon assets, and further promote the country to establish platforms for registration, verification and transaction of carbon and energy reduction, so that the fund's efforts will serve as pilot activates for making the country's carbon reduction standard, transparent, and measurable.

Forth, as for promoting the advancement of climate technologies, the fund will support independent research and development for technological breakthroughs in key industries, areas and links by various means, and encourage the research and development and dissemination of relevant technologies and methods. Besides, the fund will also fully play its role of information and exchange platform to pilot and popularize state-of-the-art technology and advanced concepts, and exchange and share information and technology through activities of systematic information gathering and filing, international communication, capacity training and etc.

Guided by the blueprint of the 12th Five-Year Plan, the fund faces bright prospects yet also enduring and arduous tasks ahead. It will grip the opportunity, forge ahead with innovation, strive for tangible results, work together with various partners from home and abroad, and play its unique and important role in facilitating China's efforts of tackling climate change and pursing low-carbon development.

Chronicle of Events

1. In Oct. 2005, the Chinese government started the preparation for establishing CDM Fund.
2. In Dec. 2005, based on the agreement between World Bank (WB) and the Chinese government, WB made a prepayment of its purchase of CERs from two Chinese CDM projects on HFC-23 decomposition, delivering 6.3 million USD of national revenue, used as start-up capital of China CDM Fund.
3. In May 2006, Asian Development Bank (ADB) provided technical assistance of 600,000 USD to China CDM Fund, to support its preparation and capacity building.
4. In Aug. 2006, the State Council approved the establishment of China CDM Fund and its Management Centre.
5. In Mar. 2007, China CDM Fund Management Center was registered at State Administration of Public Institutions Registration.
6. In Apr. 2007, the Board of China CDM Fund held the first meeting, stipulating that National Development and Reform Commission (NDRC) was chair of the Board and Ministry of Finance (MOF) as vice chair, and started the formulation of regulations and provisions of the Fund.
7. On 13th Apr. 2007, China CDM Fund received national revenue from CDM projects for the first time.
8. In Jun. 2007, the Chinese government issued *China's National Climate Change Programme*, pointing out that "to effectively use China CDM Fund".
9. On 9th Nov. 2007, MOF and NDRC jointly launched the operation of China CDM Fund.

10. In Oct. 2008, ADB provided technical assistance of 800,000 USD to support capacity building and business operation in the beginning stage of the Fund.
11. In Nov. 2008, both NDRC and MOF approved the first grants provided by China CDM Fund to support China's pressing work on tackling climate change.
12. In Mar. 2009, MOF issued *Notice on Income Tax Policies Issues of China CDM Fund and CDM Implementing Enterprises ([2009]No.30)*, alleviating tax burden for the Fund and CDM enterprises.
13. In Apr. 2009, China CDM Fund Strategic Development Committee was set up and held the first meeting.
14. On 14th Sep. 2010, upon the approval of the State Council, seven ministries including MOF and NDRC jointly promulgated *Measures for Management of China Clean Development Mechanism Fund.*
15. In Nov. 2010, China CDM Fund cooperated with China Zheshang Bank, creating a new model—clean development wealth management.
16. In Apr. 2011, the first clean development concessional loans of China CDM Fund were approved, signaling a full start of investment business of the Fund.
17. On 7th Dec. 2011, China CDM Fund and Shaanxi provincial government held a ceremony to sign a strategic cooperation agreement, showing that the Fund elevates its support to low carbon development from project level to provincial level.
18. On 1st Dec. 2011, *12th Five Year Plan on Greenhouse Gas(GHG) Control was* published, stipulating that China should make best use of China CDM Fund, diversify investment and financing channels, steer social and foreign capital to low carbon technology research and development and industrial growth and GHG control projects.
19. On 23rd Dec. 2011, the opening ceremony of the establishment of Shanghai Environment and Energy Exchange was held, with China CDM Fund as one of the largest shareholder. This marks the start the Fund's equity investment, as well as its participation in domestic carbon market planning and development.
20. On 31st Dec. 2011, national revenue from CDM projects, the major funding source of China CDM Fund, reached 10 billion RMB.

References

[1] IPCC. *Climate Change Report* 2007[M]. Sweden: IPCC Press, 2008.

[2] Zhuang Guiyang, ChenYing. *China and the Global Climate System*[M]. Beijing: World Affairs Press, 2005.

[3] United Nations. UNFCCC Berlin Mandate, 1995.

[4] United Nations. UNFCCC Kyoto Protocol, 1998.

[5] Climate Change Department of NDRC. *Clean Development Mechanism*[M]. Beijing: Standards Press of China, 2008.

[6] State Council. Circular on establishing National leading group on climate change and energy conservation and emission reduction. 2007.

[7] China CDM Fund Management Center. *Climate Change Financing*[M]. Beijing: Economic Science Press, 2011.

[8] Xie Fei, Meng Xiangming, Hu Ye. CDM: Leveraging Low Carbon Economy in Developing Countries[N]. *China Finance and Economic News*, 21st Jan. 2010 (P4).

[9] Impetuses and Distortions in China CDM. 2009. http://news.sina.com.cn/c/sd/2009-12-23/140519322122.shtml.

[10] NDRC. 2010. Circular on Low Carbon Pilot Programs in Certain Provinces and Cities.

[11] The 12th Five-Year Plan of PRC. 2011. http://news.xinhuanet.com/politics/2011-03/16/c_121193916_12.htm.

[12] Meng Xiangming, Feng Chao, Xie Fei. Global CDM Market Development and Challenges[J]. *Economic Research*. 2009(17) issue.

[13] Xie Fei, Meng Xiangming, Liu Miao. Global Carbon Market: Looking to

Break a Bottleneck[N]. *China Finance and Economic News*, 24th Jun. 2010 (P4).

[14] Carbon Finance at World Bank. State and Trends of the Carbon Market 2011.

[15] Xie Fei, Xu Mingzhu, Meng Xiangming. EU's Latest Climate Change Strategy[N]. *China Finance and Economic News*, 8th Apr. 2010 (P4).

[16] Meng Xiangming, Li Chunyi, Xie Fei. Carbon Market and CDM will not Disappear after 2012[N]. *China Energy News*, 24th Jan. 2011 (P6).